Smart Agriculture

Smart Agriculture

Emerging Pedagogies of Deep Learning, Machine Learning and Internet of Things

Edited by

Govind Singh Patel

ASSOCIATE EDITORS

AMRITA RAI, NRIPENDRA NARAYAN DAS, R. P. SINGH

CRC Press
Taylor & Francis Group
Boca Raton London New York

CRC Press is an imprint of the
Taylor & Francis Group, an **informa** business

CRC Press/Balkema is an imprint of the Taylor & Francis Group, an informa business

© 2021 Taylor & Francis Group, London, UK

Typeset by codeMantra

Library of Congress Cataloging-in-Publication
Names: Patel, Govind Singh, editor.
Title: Smart agriculture : emerging pedagogies of deep learning, machine learning and Internet of Things / edited by Govind Singh Patel, LPU Phagwara, India, Amrita Rai, UPTU, India, Nripendra Narayan Das, Manipal University Jaipur, India, R.P. Singh, Haramaya University, Diredawa, Ethiopia.
Description: First edition. | Boca Raton : CRC Press/Balkema/ Taylor & Francis Group, [2021] | Summary: "This book endeavours to highlight the untapped potential of Smart Agriculture for the innovation and expansion of the agriculture sector. The sector shall make incremental progress as it learns from associations between data over time through Artificial Intelligence, deep learning and Internet of Things applications. The farming industry and smart agriculture develop from the stringent limits imposed by a farm's location, which in turn has a series of related effects with respect to supply chain management, food availability, biodiversity, farmers' decision-making and insurance, and environmental concerns among others. All of the above-mentioned aspects will derive substantial benefits from the implementation of a data-driven approach under the condition that the systems, tools and techniques to be used have been designed to handle the volume and variety of the data to be gathered. Contributions to this book have been solicited with the goal of uncovering the possibilities of engaging agriculture with equipped and effective profound learning algorithms. Most agricultural research centres are already adopting Internet of Things for the monitoring of a wide range of farm services, and there are significant opportunities for agriculture administration through the effective implementation of Machine Learning, Deep Learning, Big Data and IoT structures"— Provided by publisher.
Identifiers: LCCN 2020036553 (print) | LCCN 2020036554 (ebook) | ISBN 9780367535803 (hardback) | ISBN 9781003138884 (ebook) Subjects: LCSH: Artificial intelligence—Agricultural applications. | Agricultural innovations. | Machine learning. | Internet of things. | Big data. Classification: LCC S494.5.D3 S623 2021 (print) | LCC S494.5.D3 (ebook) | DDC 338.10285—dc23 LC record available at https://lccn.loc.gov/2020036553
LC ebook record available at https://lccn.loc.gov/2020036554

Published by: CRC Press/Balkema
Schipholweg 107C, 2316 XC Leiden, The Netherlands
e-mail: Pub.NL@taylorandfrancis.com

www.routledge.com – www.taylorandfrancis.com

ISBN: 978-0-367-53580-3 (hbk)
ISBN: 978-0-367-68768-7 (pbk)
ISBN: 978-1-003-13888-4 (ebk)

DOI: 10.1201/b22627
DOI: https://doi.org/10.1201/b22627

Contents

List of abbreviations

ACF	Autocorrelation function
AIC	Akaike information criterion
AICc	Akaike information criterion, corrected
ANN	Artificial neural network
ARIMA	Auto-regressive integrated moving average model
ARMA	Auto-regressive moving average model
BIC	Bayesian information criterion
BPNN	Back propagation neural network
CFBP	Cascade forward back propagation network
CNN	Convolution neural network
DL	Deep learning
IoT	Internet of Things
MLR	Multiple linear regression
PACF	Partial autocorrelation function
PLL	Phase locked loop
RNN	Recurrent neural network
SOM	Self-organizing map
SVM	Support vector machine

List of symbols

a	angle AM amplitude modulation
b	angle
c	angle CORDIC coordinate transformation
DAC	digital-to-analog converter
dB	decibel
dBC	dB referred to carrier
dBm	dB over 1 mW
er	error
f	phase v radial frequency
f_i	input frequency
fm	modulating frequency
FM	frequency modulation
F_o	output frequency
F_{ref}	reference frequency
F_s	sampling frequency
G_{cd}	greatest common divisor
$H(x)$	transfer function
j	damping factor
m	index of modulation
v_n	natural frequency
v_0	center frequency

Chapter 1

Machine learning and deep learning in agriculture

Sumit Koul

LOVELY PROFESSIONAL UNIVERSITY

1.1 Introduction

Agriculture is the backbone of our economy, as the requirement of foodstuff due to rise in population is constantly increasing. There is a huge requirement of advancements in the agriculture sector such as to make precise calculations regarding yield production, using best and latest farming equipment, in order to meet the increasing needs of crop. Because of these advancements modern farming is also referred as digital farming, precise farming, intelligent farming, intensive farming, continual farming, organic farming and agribusiness. Balasankari and Salokhe (1999) studied how farmers utilize tractors in the field.

First, precision agriculture is a farming supervision theory that comprises detecting, computing and reacting to inconsistencies within the same field and other field yields. The main objective of precision agriculture study is to provide a judgment support system for managing the entire field farming with the aim to optimize profits on inputs along with the preservation of resources. Predicting weather and effect of different fertilizers with the help of remote sensing and sensors for crop health are the initial steps of precise farming. Suprem *et al.* (2013) reviewed the new technologies emerging in the agriculture sector to enhance the productivity.

Second, the agribusiness is the professional term related to agricultural yields. It is a hybrid of business in agriculture that consists of breeding, yield production, agrichemicals, farm appliances and seed supply and also the strategy of marketing and distribution. The representatives and organizations that effect food and fibre chain are the part of this agribusiness structure. Wang *et al.* (2006) introduced wireless sensors in agriculture and food industry.

Another significant aspect of modern agriculture is handling the problems in terms of yield, atmosphere impact, food safety and sustainability in the prevailing circumstances. As the global requirement of crop is increasing rapidly, crop production must be increased along with its timely availability and high nutritional quality. This can be achieved by protecting the natural ecosystem using sustainable farming practices. Farming management concept is based on observing, measuring and responding to inter- and intra-field variability in crops.

In addition to these aspects of the current agriculture industry the field of agriculture faces various problems, for example, inappropriate treatment of farms, different ailments prevailing in animals, pest infection, irregular irrigation, etc. All these problems lead to a severe damage to the yield and also prove to be hazardous to the ecosystem due to incorporation of too many chemicals in it. It is not possible to give a generalized solution for all the problems. In order to address these problems, the composite intermittent agricultural ecology should be dealt with through instantaneous observation and investigation regarding all aspects and occurrences. A remedy to this condition is possible using artificial intelligence in general and machine learning in particular. Machine learning can facilitate the agriculturalists with information to increase the crop production and diminish the initial cost as well as to balance the loss occurring during the natural calamities. Gomes and Leta (2012) reviewed a new technique of computer application in agriculture and food sector to increase the product quality. Also, Davies (2009) reviewed machine vision and its application in food and agriculture sector.

1.2 Machine learning

Machine learning is a multidisciplinary field which is a combination of computer science and statistics, where it is mostly used for analysis and classification and performs the tasks that generally humans do. For that, we need to train the computer to solve real life problems with utmost accuracy. Machine learning can be used in different situations such as face analysis, computer games, information retrieval, stock market forecasting, computational biology, DNA microarray analysis for cancer classification, epileptic seizure detection, data centre optimization, automated text categorization, medical application and analysis of gene expression data. Roberto *et al.* (2008) deduced the face recognition algorithm with the help of machine learning. Figure 1.1 represents working model of machine learning.

1.2.1 Supervised learning

As the name specifies, supervised learning means that the presence of supervisor is required to perform the task. Generally, the machine is trained by using the collected

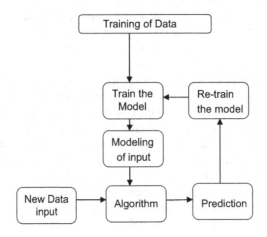

Figure 1.1 Working model of machine learning.

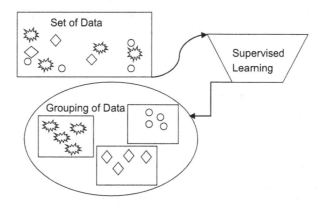

Figure 1.2 Working model of supervised learning.

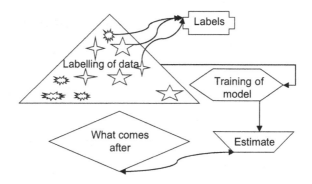

Figure 1.3 Working model of classification of supervised learning.

data which is well labelled and is known as labelled data, so that supervised learning algorithm can analyse the training data and give correct outcome using labelled data. The algorithm used in the working of supervised learning is shown in Figure 1.2.

Supervised learning is divided into two categories: regression and classification.

Regression is a technique which is used to find the relation between independent and dependent variables. Classification is the process of dividing the data into specific and distinct class where we assign a label to each class. The working model of classification supervised learning is given in Figure 1.3.

1.2.2 Unsupervised learning

In this type of machine learning there is no need to supervise the model, instead the work is initiated on its own information, where it mainly deals with unlabelled data. Unsupervised learning algorithm gives better output while performing the complex tasks compared with supervised learning. It is more unpredictable and also it helps to find the unknown pattern in data. Mehta *et al.* (2015) studied the unsupervised machine learning algorithms on precision agricultural data. Figure 1.4 represents the unsupervised learning. Unsupervised learning can be categorized into clustering and association.

1.2.3 Reinforcement learning

In reinforcement learning the agent has the ability to interact with the environment and find a better output. For this, it follows hit and trail formulae. This learning is used when there is no proper way to perform a task, but model needs to follow some strict rules to perform its duty. In this type of learning no labels are required. It has two types: one is positive and the other is negative. A survey on reinforcement learning was done by Kaelbling *et al.* (1996). Working model of reinforcement learning is shown in Figure 1.5.

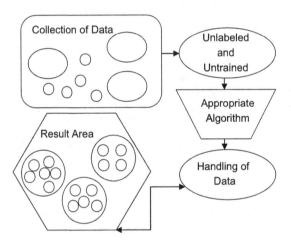

Figure 1.4 Working model of unsupervised learning.

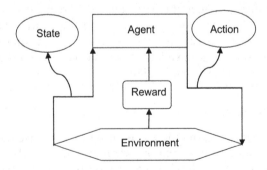

Figure 1.5 Working model of reinforcement learning.

1.3 Machine learning algorithms

As the machine learning is growing day by day, many authors are using various algorithms of machine learning to solve complex problems. Machine learning is the current research field in which a lot of research is being conducted. Figure 1.6 shows the different machine learning algorithms. Some of them are briefly explained.

1.3.1 Artificial neural network

Artificial neural network (ANN) is formed using a large number of elements known as neurons, where each neuron takes simple decisions and passes those decisions to other neurons. All the neurons are interconnected to each other, where the interconnection between these neurons is known as network function. A shallow neural network has three layers: input layer, hidden layer and output layer. Layer is the processing element of a subgroup; input layer is the first layer, and output layer is the last layer. There are many layers between the input and output layers known as hidden layers. Unsupervised ANN is also known as self-organizing map. In this network only input layers are provided, where the network needs to develop its own input incentives, so by self-organization one can collect all the input data and find the features inherent of the problem.

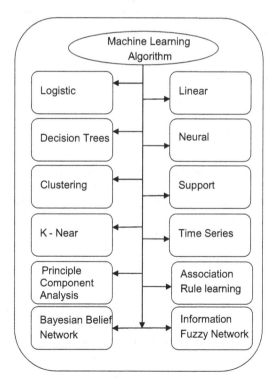

Figure 1.6 Different machine learning algorithms.

1.3.2 Support vector machine

Support vector machine (SVM) is a generally used algorithm for finding the solutions of classification and regression problems. The main aim of SVM algorithm is to create the decision boundary that can divide n-dimensional space into class, so that we can classify new data in the correct category in future. The decision boundary is known as hyper plane. SVM algorithm is used to detect the face, for classification of images, for text categorization, etc. Linear and nonlinear SVMs are types of SVM algorithm.

Consider a cat which has some features like a dog, so to determine whether it is a dog or cat, a model is created using SVM algorithm. The first step of the model is to identify images of cats and dogs so that the model can learn about different features of cats and dogs. Then it tests the odd creature, where the model creates a decision boundary between cat and dog and chooses an extreme case; that is, it will consider the extreme case of dog or cat by support vector.

1.3.3 Clustering

Clustering is dividing data into groups such that every group has similar data. It is basically a collection of data on the basis of similarities and dissimilarities between them. Clustering is important since it regulates the essential grouping among the unlabelled data. Where there is no measure for good clustering, it depends on in what way the users use the data. Density-based method, hierarchical-based methods, portioning methods and grid-based methods are the clustering methods. Density-based methods consider the cluster as the compressed region having some similarity as differentiated from the compressed region of the rest of the space. These methods have high accuracy and capacity to unite two clusters. Hierarchical-based methods form a tree-type structure based on hierarchy. New clusters are formed using earlier ones. These methods are divided into two categories: agglomerative and divisive.

1.3.4 Decision tree

Decision tree analysis is a wide-range, analytical modelling tool that has applications in different areas. Decision trees are built via an algorithm approach that identifies to split the dataset based on different conditions. In decision trees splitting of data should be continuous according to certain factors. This algorithm is described by two factors called decision nodes and leaves. Decision nodes are the nodes where data are split and leaves are the final outcomes. Classification trees and regression trees are the main types of decision trees. The main goal of the algorithm is to create a model that predicts the target variable by simple rules.

1.3.5 Principle component analysis

Principle component analysis converts correlated variables into uncorrelated variables using orthogonal transformation in a statistical procedure. Principle component analysis is used to study the interrelation between a set of variables. This algorithm is used to consider a large dataset of interconnected variables and chooses the set which best suits a model. This type of concentration of variables is known as dimensionality reduction. This method helps to reduce the complication of sets of variables. This analysis is also known as general factor analysis.

1.4 Applications of machine learning in agriculture

Some of the applications used in agriculture sectors are

- Yield prediction
- Pest and diseases detection
- Weed detection
- Soil management
- How to recognize a plant
- How to manage quality of crop
- Management of irrigation
- Welfare of animals
- Forecasting livestock

1.4.1 Yield prediction

There are many factors through which a farmer can get optimum results in agriculture. One of these factors is to predict the yield of crop. This factor includes the fertility of soil, irrigation process, climate conditions and controlling of pests. If the farmer does not follow these four factors correctly during farming, there is a huge risk of damaging of crop. Let us see some of the machine learning models which are used in the agriculture sector.

- Machine learning application helps to count the number of coffee seeds in a branch and also it segregates the coffee fruits in three categories of harvest, non-harvest and seeds with disregarded maturation stage. Also, we can estimate the weight of seeds and maturation percentage of coffee seeds. Ramos *et al.* (2017) showed how to count coffee fruits automatically from coffee tree using machine vision system (MVS). When the development of crop was going on, during the initial stage and when the harvesting was not done, using MVS technique they proved that estimation of seed count will not be too high or too low and by this they had shown that it obtains higher correlation value of 0.90.
- Amatya *et al.* (2016) developed an MVS which automatically shakes the trees and catches the cherry fruits during harvesting stage and it also detects the occluded branches and cherries which are not clearly visible. In contrast, during cherry harvesting more labour is required, which take around 50% of its annual production cost. To reduce this cost, mechanized harvesting technologies have been used such as limb-actuators that vibrate the cherry fruits so that they can be released from the branches.

This tool has generated a new era in the horticulture sector, as it has higher efficiency and is economical for the farmers. Farmers can use this technique in their agricultural work to increase the productivity.

1.4.2 Pest and disease detection

Pest and disease control is one of the main problems in today's agriculture. One of the methods to control diseases and pests is to uniformly spray the pesticides over the crops, which requires high efficiency but is not economical, and it also poses the risk of side effects such as ground water contamination and adverse impact on wildlife and ecosystem.

- Ebrahimi *et al.* (2017) developed a machine which helps to identify parasites in the green house environment through image processing. SVM method can be used for classification and targeting of parasites. The image processing methodology and SVM method having appropriate option of province and colour index proved to be successful for detection of objective with high efficiency.
- Moshou *et al.* (2014) developed an effortless and economical optical gadget for remote ailment exposure, based on awning reflectance in numerous wavebands. They investigated the difference between healthy and ailing plants in early stages of yellow rust ailment, in field images that were obtained by placing a spectrograph at spray resonant point. Then, using intensity normalization we can decrease the spectral high variability caused by canopy architecture at different illumination levels. Quadratic discriminating model based on the reflectance of these wavebands classifies healthy and disease spectra with high success rate.

1.4.3 Weed detection

For a good yield, prevention of weeds is one of the major tasks. Weed detection and prevention is difficult to discriminate from crops, so machine learning using sensors is used. This technique leads to precise detection and prevention of weeds with less expenditure and also it does not harm the environment.

- Pantazi *et al.* (2017) used remote sensing for discrimination of species and for operational weed mapping. Exposure and mapping of *Silybum marianum* weed patches by means of ordered self-organizing map is reported using a multivisionary camera which gives high-resolution images carried by Unmanned Aircraft System (UAS).

1.4.4 Soil management

The soil management plays a key role in yield efficiency, ecological stability and human health both directly and indirectly. Soil is a diverse natural resource having complex processes and fuzzy mechanism in which the temperature of soil also plays an important role in the precise investigation of climatic variations of an area and its ecological behaviour. Machine learning algorithms play a significant role in measuring soil temperature and dampness so as to understand the dynamics of ecosystems and its impact on agriculture.

- Ghosh and Koley (2014) introduced a new technique called back propagation network which gives better results for finding the good properties of soil instead of using traditional method called multivariate regression model. The working rule of back propagation networks is to train the particular crop which has certain properties.

1.4.5 Recognizing plant

In comparison with the conventional approach for classification of plant using comparison of shape and colour of leaves machine learning can give exact and faster results by analysing the leaf vein morphology which provides additional information about characteristics of leaf. The foremost objective is the automatic recognition and categorization of different plant varieties so as to evade the human expertise and also to minimize the categorization time.

- Grinblat *et al.* (2016) used deep convolution network for the problem of plant identification using leaf vein patterns. They considered three legume species of white bean, red bean and soya bean leaf vein patterns, where vein morphology was used to get the information of the leaf. It is one of the major tools for plant identification in comparison with colour and shape.
- Weiss *et al.* (2010) modelled a methodology to differentiate the species of plant using a low-resolution three-dimensional lidar sensor. The authors have modelled a feature set having common statistical features being independent of plant size. The classifiers have been trained and compared in this model with the feature set that shows high efficiency in identification.

1.4.6 Management of quality of crop

To increase the value of crop and reduce the wastage one has to classify quality of crop with minimum error. The penultimate sub-category for the crop is developed for the identification of characteristics associated with the crop class.

- Zhang *et al.* (2017) developed a model for detection and classification of the botanical and non-botanical foreign material rooted within the cotton lint at the time of harvesting process.

1.4.7 Management of irrigation

Irrigation is an important part of agriculture. It plays a significant role in yield productivity. Irrigation should neither be in excess nor less but should be balanced. To maintain these conditions certain factors need to be considered which are soil type, land topography, weather, type of crop, water quality etc.

- Hinnell *et al.*'s (2010) neuro drip is an Excel-based ANN algorithm designed to provide rapid illustration of soil wetting patterns from surface drip irrigation emitters.

1.4.8 Welfare of animals

The field of animal welfare takes care of the health and well-being of animals so as to maintain a balance in the ecosystem. The key application of machine learning is in monitoring animal behaviour during the early exposure of infection.

- Dutta *et al.* (2015) followed a two-stage machine learning framework which is an effective method for classification of cattle behaviour. Cattle sensor technology and assemble classifiers are used in the current approach to categorize and examine the behavioural changes in cattle for improving their feed.
- Pegorini *et al.* (2015) proposed a technique based on data collected by optical fibre Bragg grating sensors that are projected by machine learning technique (pattern classification). In this study, they have considered chewing process and food intake of dietary supplement. Furthermore, two more factors of hay and ryegrass that are ruminative and idleness for ingestion behaviour were considered. They showed that pattern classification differentiates the five patterns involved in chewing process.

1.4.9 Forecasting of livestock

Farm animal production deals with the problem of production system. The foremost scope of machine learning applications in farm animal production is precise judgment of monetary balances with the help of which the producers can get information based on production line monitoring and thus can gain profits. This is because the machine learning algorithms have the potential for early detection and warning of problems that give the prior information to the producers.

- Morales *et al.* (2016) studied how to detect the egg production. SVM is used for the recognition of problems in production of eggs with the help of egg production data of egg laying hens. By using optical parameter configuration of an SVM model an alert can be indicated a day in advance that can be helpful for preventive diagnosis of clinical symptoms.
- Craninx *et al.* (2008) introduced the machine learning algorithm for forecasting the rumen fermentation pattern from milk fatty acids. Machining the prediction precision of the given model is done with regression model which is dependent on both odd and multi-sequence milk fatty acids.

1.5 Deep learning

Deep learning is a modern method that has been successfully applied in various domains. Deep learning has various applications such as image processing and text classification. Since the successful rate of deep learning is very high in other domains, so it is applied to agriculture methods too. Deep learning covers several layers of neural networks designed to perform more cultured tasks. Some of the deep learning models provide remarkable results, and in terms of scale they are not matched with humans. Each layer uses the outcome of previous result as input and whole network is trained as a single chain. Deep learning platform is a platform which helps users to build different deep learning architectures or facilitate users to apply deep learning to a wide range of business applications with apps and services. One of the main differences between machine learning and deep learning is deep learning requires more data for classification whereas small data are enough for classification in machine learning. The most popular deep learning tools are theno, kera, tensor-flow, py-torch and tool-box.

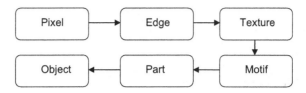

Figure 1.7 Working model for image processing.

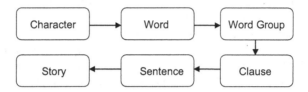

Figure 1.8 Working model for text classification.

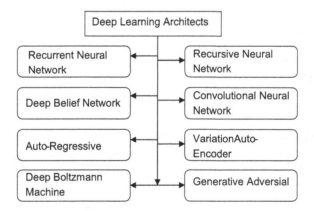

Figure 1.9 Deep learning networks.

Examples of deep learning are image processing and text classification. Figures 1.7 and 1.8 show working model of image and text classification.

Some of the deep learning architects are shown in Figure 1.9.

1.5.1 Convolution neural networks

Convolution neural network (CNN) structure is based on feed forward neural network and it is designed on an animal cortex and uses multi-layered perceptron for this process. In CNN the minimum amount of pre-processing rectified linear unit activation functions are often used. General applications are image/video recognition, natural language processing, chess etc. Convolution is used to find the features which are similar by using different places of images. It is conducted using learnable filters which are passed through the input data/images. The technique used to increase the dataset and improve CNN accuracy is known as data augmentation. Provided that large accept-able big dataset, CNN increases the exactness of correct classification. Some of the applications of ANN are decoding facial reorganization, document analysing, historic and environment collection, understanding of climate, advertising etc.

1.5.2 Recurrent neural network

Recurrent neural network (RNN) is a type of neural network where output of the previ-ous loop is considered as input for the current loop. General applications of generative neural network are speech recognition, handwriting recognition, analysis of sequence of data etc. Also, generative neural network automatically generates programming codes that give a predefined objective. Working process of RNN consists of providing input to the model. Representation of the data in the input layer is computed and sent to the hidden layer, where it conducts sequence modelling and training in forward or backward directions. Multiple hidden layers can also be used, however final hidden layer sends the processed result to the output layer. Long-short-term memory RNN is currently a popular RNN model. It is effective on data sequence that requires memory or details of last events. Some of the applications of RNN are language modelling and prediction, speech recognition, machine translation, image recognition and transla-tion. Long-short-term memory is the latest improvement of RNN network; these net-works are known as cells. These cells consider the input from previous state as present input and also decide which information needs to be considered and which one to be neglected. The previous condition, present memory and the current input combines together to predict the next output. How the RNN works is shown in Figure 1.10.

1.5.3 Generative adversarial networks

Novelty of generative adversarial networks (GAN) lies in technicality of its design. It is a type of unsupervised machine learning which includes computerized innovation such that to understand the similarities or prototype data in the manner that system produces the result. GANs are smart models to build a productive system by mod-elling a problem having two sub-models as a part of supervised learning. They are generative systems that can be educated to create illustration. GANs are the stimulat-ing and quickly rising arena. They work due to their potential of generative practical model. The vicinity of system domain is related to picture to picture conversion jobs, for example converting pictures of one season to another (e.g. summer scenes to winter scenes) or day images to night ones, in creating photo-realistic images of items, scenes and individuals which the individuals recognize as forged.

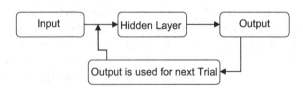

Figure 1.10 Working model of RNN model.

1.6 Application of deep learning in agriculture

Deep learning has transformed agriculture sector to its new level. Deep learning uses techniques such as conventional neural network, RNN and GAN. This gives better results and encourages agriculture domain. The procedure of deep learning uses processing of images and studying the information with efficient results. The intense growth in deep learning field has shown a lot of good results in general but it is emerging as a boon in agriculture field. By comparing deep learning with one prevailing common procedure, it can be said that this method is outperforming the existing commonly used image processing technique and has greater precision. Deep leaning allows mathematical systems which are made up of many processing stages to represent information with many levels of abstraction. Neural network and back propagation are the basis of deep learning.

1.6.1 CNN

CNN is broadly used in agriculture since it has strong capacity for image processing. Major applications of deep learning in agriculture can be classified as plant or crop classification, pest and yield prediction, robot harvesting, monitoring of disaster etc. Mainly disease recognition model for plants can be applied by leaf image and pattern classification. Berkley Vision and Learning Centre has developed a novel deep learning framework to build a disease detection model. This system is capable of identifying around 10–15 cases of ailment leaves from healthy leaves, it also has a capability to separate plant leaves from the surroundings. In 2007, to control and identify weeds an approach was designed which is a combination of CNN and K-mean features learning. Manual model for weed detection leads to false recognition and weak extraction skill in feature extraction. One of the pretrained CNN architects broadly used in classification of plants is Alexnet. Based on experimental results of Alexnet, we can say that CNN architect outperforms the machine learning algorithm, that is, hand-crafted structures for the unfairness of phonological phases. For optical image division and subsequent restoration of missing information in a time series of satellite imagery, self-organizing kohonen maps are used. In this method for post-processing setup, geospatial analysis and several filtering algorithms are used. Although CNN has several uses, it faces many challenges that have slowed down its application in plant classification. For example, each pixel of space borne SAR imagery is characterized by backscatter phase and intensity in multiple polarizations. For yield prediction and robot harvesting, fruit counting is one of the important factors. We cannot produce satisfactory results through traditional counting or video or camera image counting and also these processes are time consuming. Preprocessing of these types of images is challenging because of occlusion and illumination. Hansen *et al.* (2018) introduced a technique to identify the livestock animal such as pig using the face recognition feature of CNNs. Conventionally, radio frequency identification tags were used for detecting the animals which was earlier a cumbersome job.

To accompany a fully convolution network, a method known as blob detection was proposed. The first step is to gather the human formed labels from a set of fruit images and then this model is trained for an image segmentation performance. Then CNN is used to count the bifurcated pictures and give an approximation of number of fruits.

The last stage of the work involves applying a regression equation to map intermediate fruit count estimation to final human generated label count. Accuracy as well as efficiency is increased by combining deep learning with blob detection.

Land classification method is used to identify the land as land use and cover, for disaster assessment of risk and for food and agriculture. Overall idea of deep learning method is to integrate information developed by multiple heterogeneous sources using machine learning techniques to provide information processing and picturing capability. This process includes four steps: (i) noise filtration and data clustering, (ii) clearing land cover, (iii) map post-processing, (iv) geospatial analysis. Kussul *et al.* (2017) introduced a multilevel deep learning approach for land cover and crop types classification using multitemporal multisource satellite imaginary.

Nowadays, new technologies are emerging in agriculture sector; for example, for high-quality image processing unmanned aerial vehicles are used. For farmers, operating highly autonomous machines is quite difficult. It is not easy to operate these machines without supervision. So, detection of real time risk with these machines automatically with high reliability becomes a requirement. For sustainable land use certain conditions need to be considered, they are planning regarding reducing CO_2 emission, diminishing land degradation and improving economic returns using valuable data from satellites. For decision-making in precision agriculture and agro-industry, CNN and genetic algorithm have become convenient methods with the use of translating satellite images. Weather forecasting is one of the main factors for farmers that can be predicted using CNN. Similarly, crop yield estimation is one of the main factors for farmers, consumers as well as for the government which needs to be predicted before the harvest of the crop. CNN is not only used for crop estimation or agriculture purpose but also is used in classifying the animal behaviour.

1.6.2 RNN

Land cover classification is the key challenging area in agriculture, it involves recognizing the type and quality of land. In the past, a lot of applications were based on mono-temporal observation. Mono-temporal methods are dependent on some factors like weather.

To solve the problems related to RNN, a model known as NARX is introduced, where NARX means nonlinear autoregressive model process with exogenous input. In this method, previous prediction values are considered as input, and present and previous values are considered as exogenous input. The system not only judges the independent inputs but also the previous response of the system which makes the system more powerful. Using NARX model another model, NARXNN, is developed for estimation of time series of leaf area index (LAI). Kurumatani (2018) proposed a technique to forecast the price of agricultural product using RNN.

RNN is also used to predict the weather. Biswas *et al.* (2014) designed three models for prediction of weather: nonlinear autoregressive with exogenous inputs neural network (NARX NN), case-based reasoning model and segment case-based reasoning model. Palangpour *et al.* (2016) produced a model to identify the location of animals in the forest. In this model, particle swarm optimization algorithm combined with RNN model was used, the results obtained by this model contain less errors.

1.6.3 GAN

GAN is considered as one of the most useful neural networks in many fields. Mainly GAN is used to find the feature loss in image processing caused by down sampling. When the image is compressed, some of the information may get lost or quality of that image is lost, so we may need to recover all the original details. For this recovery, a perpetual loss function comprising adversial loss and content loss is defined. This function is then compared with the widely used pixel-wise mean squared error (MSE) loss. While working on a large number of images, this model is able to improve the quality of highly compressed images. This becomes important with all models which contain image processing work, mainly agriculture, because certain applications are dependent on remote sensing images.

Barth *et al.* (2017) proposed a model to overcome the problems associated with big amount of data obtained in deep learning systems. In the absence of manually marked information a large quantity of data (as in deep learning model or GAN-based model) is used. This is called unsupervised cycle, or generative adversarial system, to optimize the practicality of artificial agricultural pictures. Authors have proposed 10,500 artificial, 50 empirically annotated and 225 unlabelled empirical pictures to get their model working. The hypothesis made was that there was resemblance between synthetic images and empirical images which can be enhanced qualitatively to improve the transformation of features. Because of this analysis the artificial pictures were transformed easily on local characteristics like light diffusion, colour and consistency as compared with global feature translation, which was not that good.

1.7 Advantages and disadvantages in agriculture

Generally, in machine learning it is not easy to analyse the unstructured data. For this type of data analysis, applying deep learning methods will be more useful where we can use different types of data formats to make algorithm work. To find the relation between different domains which are interdisciplinary we can use deep learning algorithm. Generally, workers get tired or irresponsible or neglect the small things, but in deep learning models that is not the case. The algorithm will perform thousands of cycles of work without any error that too in short period of time. Also, the quality of work will not be affected until and unless data input by the user have some problem. In traditional learning approach, identification of features needs to be accurate, whereas in deep learning models have ability to create new features by themselves. Generally, problem-solving in machine learning is done by dividing big tasks in small tasks and combining the results of all the small tasks for the final output, whereas in deep learning tasks are solved on end-to-end basis. Deep learning requires large amount of data or information and it is expensive to use a deep learning model. One of the major disadvantages is that we are not able to find how the analysis is done inside the model. Generally, we call it black box, but sometimes knowing the analysis algorithm is important because interpretability is necessary in some domains. Nowadays as machine learning is growing in all the domains in a dramatic way, main fear is that machine learning may take all the work of humans and may drive humans into unemployment or slavery.

1.8 Companies associated with agriculture sector

Companies associated with agriculture sector have great importance in the 21st century because any country depends upon its agriculture sector as it generates big revenue that increases the wealth of the country and hence boosts its economy. New automation techniques are used to boost agriculture sector (see Table 1.1).

Table 1.1 Describing the different companies work for agriculture sector

Company	Specialization	Location	Year	Description
Blue River Technology	Controlling weeds	Sunnyvale, CA, USA	2011	Invents robotics which can effectively remove weeds from crops. Company tested its first product by spraying it on cotton plants which contained weeds.
Harvest CROO Robotics	Harvesting crop	USA	2013	Company developed robots which can harvest and simultaneously pack the crop. It was first used in strawberry farming. In a single day it can harvest 8 acres of crop as well as reduce labourers up to 30 in number.
PEAT	Detecting pests or soil defects using machines	Berlin, Germany	2015	Company has introduced the concept of deep learning which is mend for image recognition and this technique has given birth to Plantix. Plantix is used for image detection. This technology is useful in detecting soil which cannot produce good-quality crop, i.e. it detects nutrients in the soil. It is suggested that it provides 95% precise data.
Trace Genomics	Detecting soil using machine learning	CA, USA	2015	Company has developed machine learning algorithm with Illumia investor to analyse the soil by checking its quality and potentiality. Framer can get required information about the soil to get optimal results in terms of height of crop and to protect it from damage.
Sky Squirrel Technologies, Inc.	Studying crop using drones	Halifax, Canada	2012	The technology of drone was introduced by Sky Squirrel Technologies, Inc. to safeguard the farming in vineyards. This technology helps in increasing the yield of crop by decreasing the probability of diseases and pests which required more costs to eliminate. Farmer can use this technology again and again by recording the images through computer.

(Continued)

Table 1.1 (Continued) Describing the different companies work for agriculture sector

Company	Specialization	Location	Year	Description
aWhere	Predicting weather and crop continuity through remote sensing images	CO, USA	1999	Company provides technology to forecast weather conditions, analyse crop continual and detect diseases in the farm through remote sensing images via satellites using trained algorithm of machine. User utilizes the data provided by company to study the behaviour of crop on regular basis. Agronomist can find their answers through analyses of data about temperature, precipitation and wind speed.
FarmShots	Monitoring crop health and continuity	Raleigh, NC, USA	2014	Agriculture data are obtained from remote sensing images through satellite. The main goal of FarmShots is to detect the disease and nutrient condition of plants.
Abundant Robotics	Technology to harvest apples	Hayward, CA, USA	2015	For the first time, this robotic company has delivered robots which can harvest apples. In the recent years, the company is thinking on providing solutions to complex problems which are still not handled in agriculture sector.
Ibex Automation	Robotic technology in agriculture	Wortley, United Kingdom	2016	The company has come up with a robot which is autonomous in detecting weeds from the crop. It efficiently detects weed and provides system of spraying chemical only to the crop.
Hortau, Inc.	Managing irrigation system through web	San Luis Obispo, CA, USA and Quebec, Canada	2002	Basically, this company was meant for developing wireless equipment, but in 2002 it entered into the agriculture sector to manage the irrigation system. It detects the stress in plant and provides ways for optimum growth of crop with less irrigation, less energy requirement and with less impact on environment.

References

Amatya, S., Karkee, M., Gongal, A., Zhang, Q. & Whitting, M. D. (2016). Detection of cherry tree branches with full foliage in planar architecture for automated sweet-cherry harvesting. *Biosystems Engineering*, 146, pp. 3–15.

Balasankari, P.K. & Salokhe, V.M. (1999). A case study of tractor utilization by farmers, Coimbatore district, India. *Agricultural Mechanization in Asia, Africa and Latin America*, 30, pp. 14–18.

Barth, R., Ijsselmuiden, J. M. M., Hemming, J. & Van Henten E. J. (2017). Optimising realism of synthetic agricultural images using cycle generative adversarial networks. *Proceedings of the IEEE IROS Workshop on Agricultural Robotics/Kounalakis, Tsampikos, van Evert, Frits, Ball, David Michael, Kootstra, Gert, Nalpantidis, Lazaros*, Wageningen: Wageningen University & Research, pp. 18–22. http://library.wur.nl/WebQuery/wurpubs/533105.

Biswas, S. K., Sinha, N., Purkayastha, B. & Marbaniang, L. (2014). Weather prediction by recurrent neural network dynamics. *International Journal of Intelligent Engineering Informatics*, 2, pp. 166–180.

Craninx, M., Fieveza, V., Vlaeminck, B. & De Baets, B. (2008). Artificial neural network models of the rumen fermentation pattern in dairy cattle. *Computers and Electronics in Agriculture*, 60, pp. 226–238.

Datta, R., Smith, D., Rawnsley, R., Bishop-Hurley, G., Hills, J., Timms, G. & Henry, D. (2015). Dynamic cattle behavioural classification using supervised ensemble classifiers. *Computers and Electronics in Agriculture*, 111, pp. 18–28.

Davies, E. R. (2009). The application of machine vision to food and agriculture: a review. *Imaging Science Journal*, 57, pp. 197–217.

Ebrahimi, M.A., Khoshtaghaza, M.H., Minaei, S. & Jamshidi, B. (2017). Vision-based pest detection based on SVM classification method. *Computers and Electronics in Agriculture*, 137, pp. 52–58.

Ghosh, S. & Koley, S. (2014). Machine learning for soil fertility and plant nutrient management using back propagation neural networks. *International Journal on Recent and Innovation Trends in Computing and Communication*, 2, pp. 292–297.

Gomes, J. F. S. & Leta, F. R. (2012). Applications of computer vision techniques in the agricultureand food industry: a review. *European Food Research and Technology*, 235, pp. 989–1000.

Grinblat, G. L., Lucas, C. U., Mónica, G. L. & Granitto, P. M. (2016). Deep learning for plant identification using vein morphological patterns. *Computers and Electronics in Agriculture*, 142, pp. 418–424.

Hansen, M. F., Smitha, M. L., Smitha. L. N., Michael, G., Salterb, M. G., Baxterc, E. M., Farishc, M. & Grieved, B. (2018). Towards on-farm pig face recognition using convolutional neural networks. *Computers in Industry*, 98, pp. 145–152.

Hinnell, A. C., Lazarovitch, N., Furman, A., Poulton, M. & Warrick, A. W. (2010). Neuro-Drip: estimation of subsurface wetting patterns for drip irrigation using neural networks. *Irrigation Science*, 28, pp. 535–544

Kaelbling, L. P., Littman, M. L. & Moore, A. W. (1996). Reinforcement learning: a survey. *Journal of Artificial Intelligence Research*, 4, pp. 237–285.

Kurumatani, K. (2018). Time series prediction of agricultural products price based on time alignment of recurrent neural networks. *17th IEEE International Conference on Machine Learning and Applications*, pp. 82–88.

Kussul, N., Lavreniuk, M., Skakun, S. & Shelestov, A. (2017). Deep learning classification of land cover and crop types using remote sensing data. *IEEE Geosciences and Remote Sensing Letters*, 14, pp. 778–782.

Mehta, P., Shah, H., Kori, V., Vikani, V., Shukla, S. & Shenoy, M. (2015). Survey of unsupervised machine learning algorithms on precision agricultural data. *IEEE Sponsored 2nd International Conference on Innovations in Information, Embedded and Communication Systems (ICIIECS)*, pp. 1–8.

Morales, I.R., Cebrián, D. R., & Blanco, E. F. (2016). Early warning in egg production curves from commercial hens: an SVM Approach. *Computers and Electronics in Agriculture*, 121, pp.69–179.

Moshou, D., Bravo, C., Jonathan, W., Wahlen, S., Cartney, M. A. & Ramona, H. (2014). Automatic detection of 'yellow rust' in wheat using reflectance measurements and neural networks. *Computers and Electronics in Agriculture*, 44, pp. 173–188.

Palangpour, P., Venayagamoorthy, G. K. & Duffy, K. (2006). Recurrent neural network based predictions of elephant migration in a South African game reserve. *The 2006 IEEE International Joint Conference on Neural Network Proceedings* , https://ieeexplore.ieee.org/document/1716662, pp. 4084–4088.

Pantazi, X. E., Tamouridou, A. A., Alexandridis, T. K., Lagopodi, A. L. & Kashefi, J. (2017). Evaluation of hierarchical self-organising maps for weed mapping using UAS multispectral imagery. *Computers and Electronics in Agriculture*, 139, pp.224–230.

Pegorini, V., Karam, L.Z. & Pitta, L.S.R. (2015). In vivo pattern classification of ingestive behavior in ruminants using fbg sensors and machine learning. *Sensors*, 15, pp. 28456–28471.

Ramos, P. J., Prieto, F. A., Montoya, E. C. & Oliveros, C. E. (2017). Automatic fruit count on coffee branches using computer vision. *Computers and Electronics in Agriculture*, 137, pp. 9–22.

Roberto, V., Sebe, N., Gevers, T & Cohen, I. (2008). Machine learning techniques for face analysis. Eds. Cord, M. & Cunningham, P. *Machine Learning Techniques for Multimedia*. Springer, Berlin Heidelberg, 159–187.

Suprem, A., Mahalik, N. & Kim, K. (2013). A review on application of technology systems, standards and interfaces for agriculture and food sector. *Computer Standards & Interfaces*, 35, pp. 355–364.

Wang, N., Zhang, N. & Wang, M. (2006). Wireless sensors in agriculture and food industry—recent development and future perspective, computers and electronics in agriculture. *Elsevier Science*, 50, pp. 1–14.

Weiss, U., Biber, P., Laible, S., Bohlmann, K. & Zell, A. (2010). Plant species classification using a 3D LIDAR sensor and machine learning. *Ninth International Conference on Machine Learning and Applications*, pp. 12–14.

Zhang, M., Changying, L. & Fuzeng, Y. (2017). Classification of foreign matter embedded inside cotton lint using short wave infrared (SWIR) hyperspectral transmittance imaging. *Computers and Electronics in Agriculture*, 139, pp. 75–90.

Chapter 2

Descriptive and predictive analytics of agricultural data using machine learning algorithms

R. Suguna and R. Uma Rani

SRI SARADA COLLEGE FOR WOMEN (AUTONOMOUS)

2.1 Introduction

2.1.1 Agriculture in India

India is a major country whose economy is based on agriculture. There are various crops such as fruits, vegetables, oil seeds and millets. Villages are the major source of crop in our country. Besides exporting agricultural products at global level, almost 120 countries are benefitted by India's agriculture. According to the GDP data of India, there has been an increase of 6091.05 INR billion in January 2020 from 3664.29 INR billion in July 2019 (GDP 2019). Figure 2.1 shows the GDP of India.

2.1.2 Rainfall and crop production

Rainfall is the base variable for crop production. Meteorological department estimates the rainfall level of India. There are six regional meteorological centres throughout India. India is the highest producer of crops. Crops are divided based on seasons such as rabi, kharif and zaid crops. Rabi crops are harvested during the months of September–October. Kharif crops are harvested during the months of April–May. Zaid crops are growing in the time between March and June. Maximum number of crops need water for being cultivated (Ibn Musah *et al.* 2018).

2.1.3 Data analytics

Data analytics and advanced mining technologies analyse data with exploratory data analysis which is used to describe data and establish a pattern and relationship in the data with the help of statistical methods. Confirmatory data analysis evaluates data using statistical methods. Data analytics analyse the data in the manner described in Figure 2.2.

Data can be analysed in different ways depending on whether it is past or future data. Based on different types of analytical methods, data can give fruitful solutions. The analytical methods comprise following methods (Figure 2.3):

Descriptive analytics: In this analysis, history of past data is analysed, that is how data got modified according to the process. For example, amount of rainfall occurred in various districts and other information of rainfall data are considered.

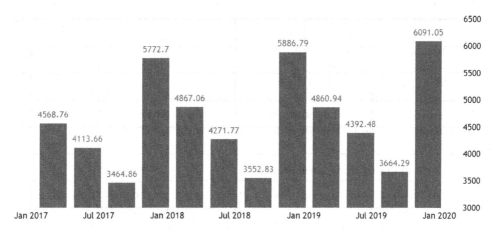

SOURCE: TRADINGECONOMICS.COM | MINISTRY OF STATISTICS AND PROGRAMME IMPLEMENTATION (MOSPI)

Figure 2.1 GDP of India.

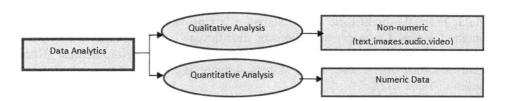

Figure 2.2 Various approaches of analysis.

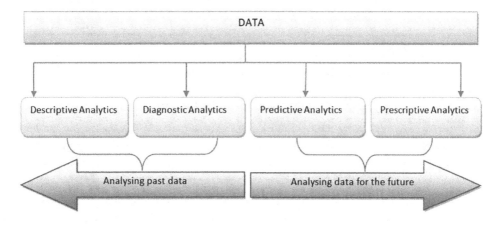

Figure 2.3 Types of data analytics.

Diagnostic analytics: In this type of analysis, the reason for problems and other special deeds according to the process are analysed. For example, analysing the causes for excessive or less rainfall at a particular location.

Predictive analytics: This is a forecasting method, which analyses the data and predicts the future level of the process. For example, in rainfall data, the amount of rainfall will be predicted based on the history of past rainfall data.

Prescriptive analytics: This is an analytical method, which optimizes the predictive result of the process. For example, increasing the accuracy and using optimization techniques to produce better outcomes in the rainfall data.

2.1.4 Life cycle of data analytics

Data analytics life cycle consists of six stages to solve problems and taking better decision according to the data. Like data mining, analytics also follows a prescribed channel to handle the data. The phases include discovering the problem, preparing data, model planning, model building, visualization of results and operationalization (Poudel 2016) (Figure 2.4).

2.1.5 Statistical models as machine learning techniques

Statistics is a branch of mathematics based on mathematical techniques applied to data. From the prehistoric times, statistics has provided efficient problem-solving methodologies. The major operations of statistics are collecting and reviewing data, analysing and interpreting data and showing results in summarized manner (Kumar and Choudhry 2010). Some of the statistical methods useful in machine learning are given in Table 2.1.

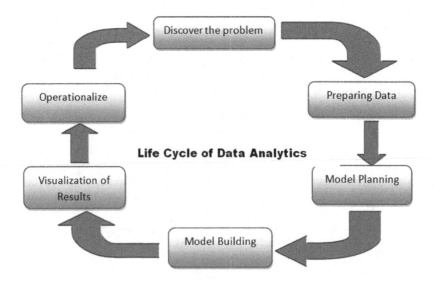

Figure 2.4 Different phases of data analytics.

Table 2.1 Statistical models

Description	Formulae	Explanation
Arithmetic mean	$\bar{x}$	Average value of the samples
Standard deviation	$s = \sqrt{\dfrac{\sum (x - X)^2}{n - 1}}$	Average distance from the mean. N is number of samples and every value x, mean value $\bar{X}$
Coefficient of variation	$c_v = \dfrac{s}{\bar{x}} \times 100$	It is measure of relative variability. S is standard deviation, $\bar{X}$ is mean
Coefficient of skewness	$sk = \dfrac{x - mo}{s}$	Mean and mode values are used and s is standard deviation.
Correlation	$r = \dfrac{1}{(n-1)} \left[\dfrac{\sum (x - \bar{x})(y - \bar{y})}{S_x S_y} \right]$	Measure of association between variables x and y.
Regression	$y = b_0 + b_1 \times x_1$	Relationship among dependent and independent variables. There are some other formulas available for regression line and slope etc.

Machine learning is an emerging area which is successor of artificial intelligence. It implies process of training a machine without explicit programming. There are various kinds of techniques used to handle various types of problems:

Supervised learning: In this kind of learning, the users have prerequisite knowledge about the output of a problem. The labelled data are evaluated with the help of an expert. Prediction can be made along with labelled data. Algorithms for the supervised learning are classification and regression.

Example: SVM, discriminant analysis, naive Bayes, K-nearest neighbour (KNN), linear regression, logistic regression, neural networks, decision trees.

Unsupervised learning: In this kind of learning, the users cannot have a prerequisite knowledge of output of problem. Unlabelled data are trained without the supervision of an expert. Algorithms for unsupervised learning are clustering.

Example: K-means, K-mediods, fuzzy c-means, hidden Markov, Gaussian mixture, hierarchical.

Reinforcement learning: This is another kind of learning method which is somewhat different from the aforementioned methods. Many disciplines such as operation research, game theory and control theory use reinforcement learning method (Daume 2012).

Example: Autonomous vehicles, games.

2.2 Literature review

A wide range of statistical methods are used for the various analyses done in agriculture sector to make today's agriculture smarter. Technology comprising models has proved to be helpful in getting more production in agriculture and it also can easily

Table 2.2 Review of works regarding agriculture

Authors	Work regarding agriculture	Description of techniques
Poudel and Shaw	Relationship between climate variability and crop yield	Regression model
Rohitashw Kumar and Harendar Raj Gowtham	Impact on climate change and crop productivity	Summary of data
Ibn Musah et al. (2018)	Climate change and variability in temperature in Ghana	Continuous distribution
Kumar et al. (2018)	Efficient crop yield estimation of sugarcane	SVM, KNN, least square SVM
Liakos et al. (2019)	Analysis of soil, water, livestock management	ANN, SVM
Priya et al.	Prediction of yield of crop	Random forest
Renuka	Analysis of sugarcane crop data of Karnataka state	KNN, SVM, decision trees
Zingade et al. (2017)	Android application for suggesting the best profitable crops for the given weather condition	Multiple linear regression
Chourasiya et al. (2019)	Seed classification for crop production	ANN, SVM, MLR
Veenadhari et al.	Prediction to control climatic parameters for Madhya Pradesh	C4.5 algorithm
Abih	Rainfall variability on Ethiopia	Regression, correlation
Kumar et al. (2019)	Comparison of weather forecasting using ML algorithms	Decision trees, KNN
Kagita et al.	Optimum agriculture land allocation of Krishna delta	Fuzzy membership functions, GA
Palanivel and Surianarayanan (2019)	Comparison of crop yield with ML algorithms	Linear regression, ANN
Kyei-Mensah et al. (2019)	Analysis of rainfall distribution and crop production for Ghana	Pearson correlation
Ndamani and Watanabe (2013)	Inter and intra-annual rainfall variability for crop production in northern Ghana	Coefficient of variation, correlation analysis
Kyada et al.	Sensitivity analysis of rainfall forecasting of Gujarat	ANN, adaptive neuro-fuzzy inference system
Sumi et al.	Machine learning methods for forecasting daily and monthly rainfall of Fukuoka in Japan	Principal component analysis (PCA), artificial neural network (ANN), multivariate adaptive regression splines (MARS),
Peprah (2014)	Correlation among temperature and rainfall of Asunafo forest, Ghana.	Correlation analysis
Bewket (2009)	Food grain production and rainfall variability for Amhara region in Ethiopia.	Spearman correlation analysis, coefficient of variation.
Shinde and Khadke (2017)	Maharastra's rainfall on crop production	Standard deviation, Pearson correlation
Varsha and Pai	Rainfall prediction of India	Fuzzy C-means clustering, FRBCS
Parmar et al. (2017)	Comparison of machine learning algorithms on rainfall prediction	ANN, SOM, CFBP, BPNN, SVM

(Continued)

Table 2.2 (Continued) Review of works regarding agriculture

Authors	Work regarding agriculture	Description of techniques
Zaman (2018)	Machine learning model on rainfall for Bangladesh	Regression, decision trees, random forest, KNN
Yousif *et al.* (2018)	Implications of rainfall variability for agricultural production in Eastern Sudan	Coefficient of variation, simple linear regression, correlation
Bala Sai Tarun *et al.* (2019)	Prediction of rainfall with machine learning algorithms	SVM, CART, GA
Refonaa *et al.*	Rainfall forecast of Chennai	Linear regression
Arvind *et al.* (2017)	Analysis of rainfall data for Trichy, Tamilnadu	Standard deviation, CV, probability distribution
Lobell *et al.*	Effects of climate change on crop yield	Regression model
Wenjiao *et al.*	Analysis of crop yields based on climatic contributions	Time series model, cross-section model, panel model
Tailor	Crop production of sugarcane in Olpad region of Gujarat	Regression, ANOVA
Khatri (2013)	Water irrigation system	Fuzzy logic and artificial intelligence
Tzimopoulos *et al.* (2018)	Relationship between rainfall and altitude of different meteorological stations in Kerala	Fuzzy linear regression
Archontoulis *et al.*	Applications and regression models in agriculture	Nonlinear regression model
Menaka and Yuvaraj	Crop yield prediction models	Adaptive Neuro-Fuzzy Inference System, MLR

handle some of unexpected problems such as floods, draughts and demands during food shortages. Table 2.2 shows the various processes carried out in agriculture sector through analytics and machine learning techniques.

Regression is a model that demonstrates the liaison between the dependent and independent variables. Most of the researchers used regression model in their studies in agriculture, while many authors concentrated on data of rainfall and crop production of various regions all around the globe. Regression is one of the supervised machine learning technique which is very flexible and provides high-quality results (Palanivel and Surianarayanan 2019). The rainfall data of different areas of Eastern Sudan have been analysed and it provided reliable R^2 values for each station. In Bangladesh, a study of various other algorithms such as naive Bayes, decision trees and regression with 77% of accuracy of results (Zaman 2018) was conducted.

Correlation is another technique of machine learning which is used to evaluate the strapping of two or more variables. In the study of rainfall variability and crop production conducted in Ghana (Ndamani and Watanabe 2013) correlation technique was used to calculate the effect of rainfall (independent variable) on crop yield (dependent variable). In the study of rainfall variability in Ethiopia (Abih 2011), correlation technique was used to state the association between the spring and summer rainfall. Peprah (2014), in the study of Asunafo forest of Ghana, shown the relationship between the climatic conditions of various crops like rice, maize, cassava, cococyam, plantain and yam and rainfall with the help of correlation.

Coefficient of variation is another technique which is used to determine of dispersion of data points in the data sequence around the mean. Ndamani and Watanabe

(2013) used coefficient of variation for determining the relationship between the annual and seasonal rainfall of Ghana. Annual rainfall shows coefficient of variation of 0.18, which is taken as moderate rainfall based on correlation and mean. In the study conducted in Ethiopia, Bewket (2009) also used coefficient of variation for comparing the annual, seasonal and daily rainfall levels. Spearman correlation was applied on rainfall data to get its significance. Least square regression was used to best fit the parameter of rainfall.

Continuous distribution is another statistical method which can handle continuous data. When there is a wide range of values, continuous distribution is applied. Some functions such as Weibull, Gumbel and Frechet are the part of extreme value distribution. These functions are used to maximize the maxima value of random variable. GEVD is also a function of continuous distribution which was used in a study done by (Ibn Musah *et al.* 2018) to get the maximum likelihood values for temperature and rainfall variables.

Probability distribution depicts various possible effects of an incident. It is a mathematical based statistical method. It is divided into continuous probability distribution and discrete probability distribution. This method was used in the study of Tiruchirappalli rainfall data, for evaluating the distribution of rainfall over the region. Mean and standard deviation values are calculated and based on that chi-square values are calculated with various distribution functions. Chi-square values are used to rank the best fit of the distribution (Arvind *et al.* 2017).

Multiple linear regression (MLR) is modified form of linear regression and it is an effective machine learning technique. Linear regression can handle one independent variable to a dependent variable. But in MLR, more than one independent variables can be assigned to a response variable. Seed classification is done based on soil dataset. Parameters of soil were assigned as independent variables and seed type as a decisive factor in the work of Chourasiya *et al.* (2019). An android application has been developed by Zingade (2017) with the machine learning technique of MLR. In this application crop prediction as done according to the environmental conditions.

SVM, artificial neural network (ANN), decision trees, random forest and KNN are some supervised machine learning algorithms. Comparative analysis was made with the algorithms such as SVM, ANN, GA and CART on rainfall data. In comparison with other algorithms, ANN provided better performance with 86% accuracy (Bala Sai Tarun *et al.* 2019). Other algorithms such as SOM, SVM, BPNN, CFBPN and ANN are discussed and reviewed for their benefits in agricultural scenario. The author reviewed various kinds of works regarding agriculture (Parmar *et al.* 2017).

2.2.1 Review of data analytics

The study of data analytics can be divided in traditional data analytics and big data analytics. The main processes are input, processing and output, and the framework of the analytics is determined on the basis of perspective-oriented and result-oriented concerns. The tools used for processing the data analytics environment are Apache, SPSS, Storm, Dryad, R, Tableau, Japer software. These are the various tools used for big data as well (Acharjya 2016).

Descriptive analytics is one of the method analytics which analyses the past data with summary of the data. In agriculture, crop yield data for 10 years is collected and descriptive analytics is applied on the data. KNN algorithms are also applied on the data. Accuracy is then measured for the KNN algorithm with root mean square error value (Renuka 2019). Predictive analytics helps to analyse the past data and forecast the future trend according to the data. Autoregressive integrated moving average (ARIMA) model is used to forecast the data and then SVM, KNN and ordinary least square are used for crop prediction (Kumar *et al.* 2018).

2.3 Methodology

In this chapter, an illustration of forecasting the rainfall level using the time series analysis method is provided. The flow of the process goes as data collection, then cleansing of the data, model building and finally visualization (Figure 2.5).

2.3.1 Data collection

The rainfall data are collected from the website of agriculture department (data.gov. in). Rainfall dataset consists of 150 years of data from 1901 to 2015. The attributes of the data such as annual rainfall and rainfall of every season for each year are provided on the website.

2.3.2 Data cleansing

The dataset contains null values in the some of the rows. To identify the null values isnull() function is used and the missing values are filled by the mean values of the rows. Applying the mean values is one of the methods used to handle the missing values.

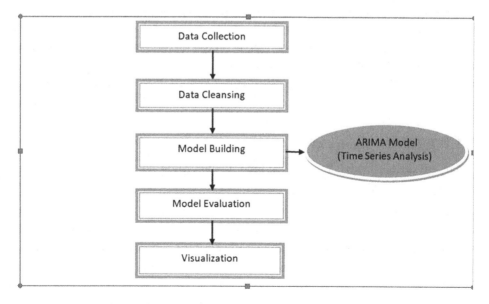

Figure 2.5 Process of ARIMA model.

2.3.3 Model planning

Descriptive analytics: Descriptive analytics is an analysis based on descriptive statistical methods such as mean, median, mode, standard deviation, variability, skewness and kurtosis. These are collectively called measures of central tendency and dispersion of data.

- Mean and median: Average value of a series of data is called mean, and the middle value of the series is called median.
- Standard deviation and variability: The sequence depends on the mean of a series. The amount of difference from the mean for each data item is called standard deviation. This is the square root of the variance. A squared difference from the mean is called variance. The variability expresses the level scattering of data throughout the series.
- IQR: The difference between the 75th percentile of series and 25th percentile of the data series is IQR. It provides the middle value, which is 50% of the data. The expression Q3–Q1. Q3 gives 75th percentile data and Q1 gives 25th percentile data.
- Skewness and kurtosis: Skewness is a representation of the data. It provides the lack of proportion in data allocation. There are two types of skewness. One is positive skewness which means the mode value is less than the mean and median values. In negative skewness, mode value is greater than the mean and median values.
  - If $-0.5 \leq$ skewness ≤ 0.5, then data are symmetrical.
  - If $-1 \leq$ skewness ≤ -0.5, then data are negatively skewed.
  - If $0.5 \leq$ skewness ≤ 1, then data are positively skewed.
- Kurtosis is the method to find the outliers in the sequence of data. This can be divided into high and low kurtosis. If the kurtosis is high, the data are having high amount of outliers. It should be considered for further processes. If the kurtosis is low, then the data are having low outliers. Based on the value of kurtosis, it can be classified into following three types:
  1. Mesokurtic: The kurtosis of the distribution is same as the normal distribution value.
  2. Leptokurtic: If kurtosis > 3, then it is high than the mesokurtic which means the data contains high outliers.
  3. Platykurtic: If kurtosis < 3, then it is meant by shorter distribution and data are having low level of outliers (Kumar and Choudhry 2010).

2.3.4 Predictive analytics

There are various models that can be used for prediction purpose. Time series is a sequence of data points which is well-ordered based on time. Time series can be expressed as

$$Y_t = f(t), \tag{2.1}$$

where Y_t is the variable's value in the study at time t. The components of time series analysis are

- Trend
- Seasonality

- Cyclic
- Randomness

Trend: It refers to increasing or decreasing data values over a long period of time. It may be classified as linear or nonlinear trend.

Seasonality: The nature of the data occurrences over a period of time. The data can be generated on weekly, monthly or yearly basis. This is known as periodic fluctuation of data. So, observation of data based on the fixed period is called seasonality of data.

Cyclic: This is also a periodic fluctuation, but it is not fixed on the seasonality component.

Randomness or uneven movements: Irregular variation of the observed values not happening in a cycle. For example: flood, wars, etc.

Time series analysis designed by George Box and Gwilym-Jenkins together is called Box–Jenkins methodology. The major processes of this method are

1. Selecting a model
2. Finding optimal parameters
3. Building ARIMA model
4. Making predictions

Selecting a model: Selection of data and checking whether the data are having a trend, seasonality or randomness.

Finding optimal parameters: The data should be stationary, which is an important feature of time series. Data should have constant mean and variance of data. If a model is having constant mean and variance of data, then it is called stationary. There are some techniques available to make a model stationary. They are

- Detrending: This is a technique to remove the trend.

$$X(t) = (mean + trend \times t) + error \qquad (2.2)$$

- Differencing: This is used to remove the non-stationarity and it is the integration process.

$$X(t) - X(t-1) = ARMA(p,q) \qquad (2.3)$$

p = AR (autoregressive)
q = MA (moving average)

Building ARIMA model: ARIMA $(p,d,q) \times (P,D,Q)$ is a model which works on stationary data. Where p is AR order, q is MA order and d is degree of differencing. If p=0, then the data are stationary data. Then ACF and PACF should be calculated. ACF is analogous to the correlation function of two variables and the limitation of ACF is −1 and 1.

$$ACF(h) = \frac{cov(y_t, y_{t+h})}{\sqrt{cov(y_{t,} y_t) cov(y_{t+h}, y_{t+h})}} = \frac{cov(h)}{cov(0)} \qquad (2.4)$$

where t is time and h=0, 1, 2, 3...

PACF is correlation among the remaining values in ARIMA. PACF is expressed as

$$PACF(h) = corr\left(y_t - y_t^*, y_{t+h} - y_{t+h}^*\right) \quad \text{for } h \geq 2$$
$$= corr\left(y_t, y_{t+1}\right) \quad \text{for } h = 1,$$

(2.5)

where $y_t^* = \beta_1 y_{(t+1)} + \beta_2 y_{(t+2...)}, y_{t+h}^* = \beta_1 y_{t+h-1} + \beta_2 y_{t+h-2...}$

Linear regression is used to eliminate the consequences of variables y_t and y_{t+h}. $h - 1$ and β values are based on linear regression (EMC 2015).

Making predictions: Using the residuals of ACF and PACF data, prediction of future points can be done with forecast() function.

2.4 Results and discussion

2.4.1 Descriptive analytics

Figure 2.6 shows the overall rainfall level of India from 1901 to 2015. Maximum and minimum rainfall values are highlighted. Maximum rainfall was 1480.3 mm, which occurred in 1917, and minimum rainfall was 920.8 mm, which occurred in 2002.

2.4.2 Nature of data: skewness and kurtosis

Figure 2.7 shows the skewness and kurtosis of the dataset. The skewness of the rainfall data is 0.01999941. It shows that the distribution of data is positively skewed. Kurtosis value is 2.763914. This is less than 3 which means the data contain low-level outliers.

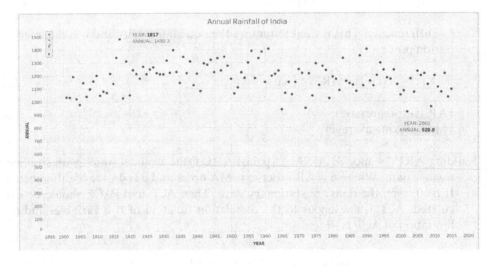

Figure 2.6 Annual rainfall level of India from 1901 to 2015.

density.default(x = s, bw = k)

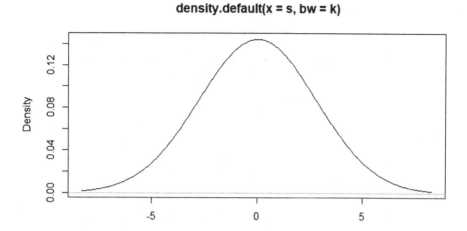

N = 1   Bandwidth = 2.764

Figure 2.7 Skewness and kurtosis of rainfall data.

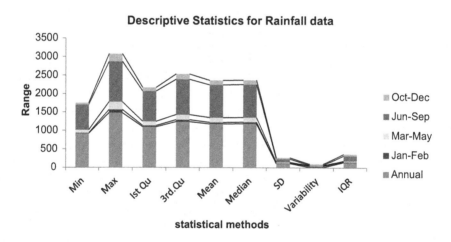

Figure 2.8 Graphical representation of descriptive parameters.

Table 2.3 explains the overall summary of data with parameters like minimum, maximum, mean, median, IQR, standard deviation values for the annual and seasonal rainfall (Figure 2.8).

Seasonal rainfall level:

The seasonal rainfall level expresses the maximum rainfall that occurred in 1960 in the October–December interval. Based on the comparison of four seasons, June–September and October–December seasons faced high rainfall (Figure 2.9).

Table 2.3 Descriptive parameters of rainfall data

	Minimum	Maximum	First quarter	Third quarter	Mean	Median	Standard deviation	Variability	IQR
Annual	920.8	1480.3	1102.4	1242.5	1182.0	1190.5	110.68	9.36	141.15
January–February	11.70	86.30	33.80	51.40	43.19	41.30	14.47	33.51	17.6
March–May	84.5	209.7	112.3	139.7	128.7	125.1	22.89	17.79	27.3
June–September	679.5	1094.5	823.5	959.6	890.3	897.8	89.17	10.01	136.1
October–December	52.7	207.50	97.05	142.15	119.88	116.20	32.46	27.08	45.1

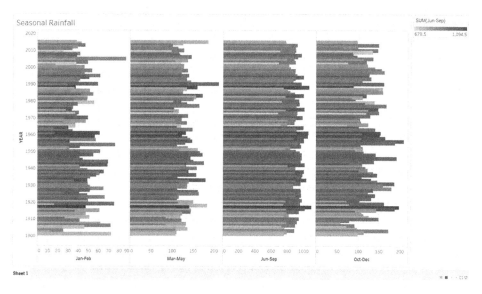

Figure 2.9 Rainfall level in seasonal data.

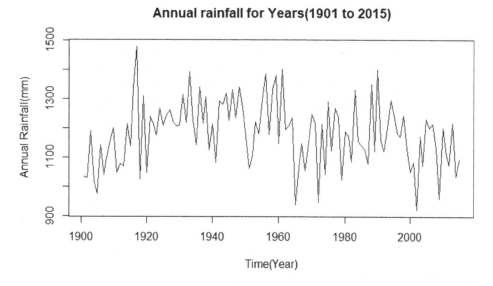

Figure 2.10 Flow of rainfall data based on time series.

2.4.2 Predictive analytics

Prediction of rainfall data was done with the help of time series analysis. ARIMA is one of the models used for time series analysis. Figure 2.10 shows the normal flow of time series.

Decomposition of additive time series

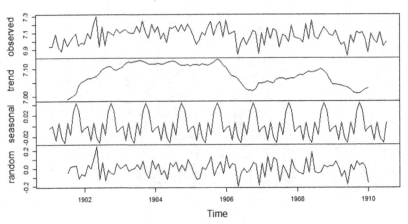

Decomposition of multiplicative time series

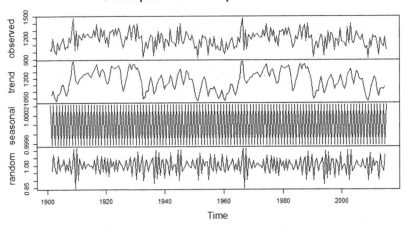

Figure 2.11   Decomposition of additive and multiplicative time series.

Figure 2.11 explains the decomposition of additive and multiplicative time series of rainfall data. To make the data stationary, the components such as trend, seasonality, randomness should be considered. If there is increasing trend, the amplitude of seasonality will also increase. But in our rainfall data, seasonality sometimes increases and sometimes decreases. In additive time series the components are added together but in multiplicative time series components are multiplied together and log of data series is taken.

2.4.4 Autocorrelation and partial autocorrelation functions

Figure 2.12 shows the autocorrelation and partial autocorrelation of rainfall data. These are the plots used to display the correlated data with the significant level. In the figure, the data are correlated within the boundary level with 95% confidence interval

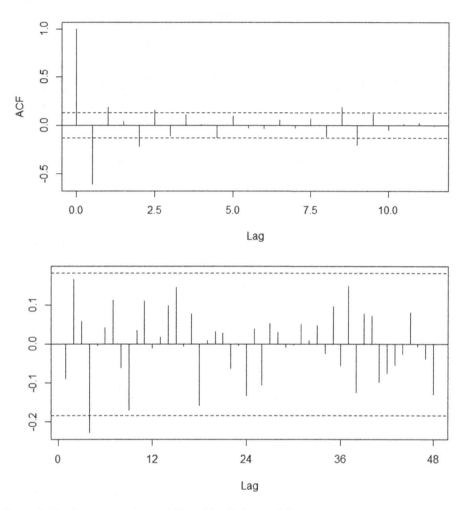

Figure 2.12   Representation of ACF and PACF for rainfall data.

significant level. Partial autocorrelation is the relationship between the observed data which has applied time series and observed values of pre-level of time series. So, in this plot most of the data lie between the significance boundaries compared with the ACF. The term lag is used in both ACF and PACF plots which is time units of t and h, where h is quantity value of covariance.

2.4.5 Measuring goodness of fit

The components AIC, AICc and BIC are the estimators of best fit model of ARIMA. These coefficients are supported to maximize the log likelihood value of ARIMA model. Rainfall data provide best fit along with $(0, 1, 1) \times (1, 0, 0)$ model and its coefficients are smaller than the other ARIMA models (Table 2.4).

Residual graph is shown in Figure 2.13 for the fitted ARIMA model on rainfall data.

Table 2.4 Fitness parameters of ARIMA model

ARIMA model (p,d,q) × (P,Q,D)	AIC	AICc	BIC
(0,1,0) × (1,0,0)	1462.92	1474.48	1474.25
(0,1,1) × (1,0,0)	**1394.94**	**1394.84**	**1384.96**
(0,1,2) × (1,0,0)	1474.22	1488.12	1474.39
(1,1,0) × (1,0,0)	1474.25	1474.42	1488.36
(1,1,1) × (1,0,0)	1462.25	1464.76	1473.25

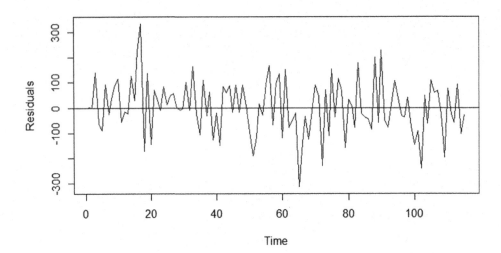

Figure 2.13   Residual plot for ARIMA model.

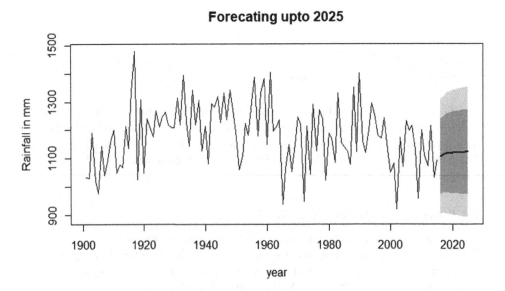

Figure 2.14   Graphical representation of forecasting.

Table 2.5 Forecasting up to 2025 based on ARIMA model

Year	Point forecast	Low 80	High 80	Low 95	High 95
2016	1103.422	969.6137	1237.231	898.7797	1308.065
2017	1113.926	979.6177	1248.235	908.5191	1319.333
2018	1115.651	974.9193	1256.383	900.4205	1330.881
2019	1117.800	976.0551	1259.546	901.0196	1334.581
2020	1118.695	974.9715	1262.418	898.8889	1338.501
2021	1120.554	975.6334	1265.474	898.9174	1342.190
2022	1118.604	972.2957	1264.912	894.8448	1342.363
2023	1119.431	971.8897	1266.972	893.7861	1345.076
2024	1119.733	970.9258	1268.541	892.1519	1347.315
2025	1118.581	968.5466	1268.615	889.1232	1348.039

2.4.6 Forecasting of ARIMA

Forecasting of rainfall from 2016 to 2025 can be done with low and high levels of expectation. The confidence levels are 80% and 95%. Figure 2.14 shows the graphical representation of forecasting (Table 2.5).

2.5 Conclusion

Data analytics can solve many problems in the fields of business, medicine and agriculture. Machine learning algorithms combined with data analytics methods provide most appropriate results for many problems. Time series analysis is the most suitable method to analyse past data. It is a vast technique. ARIMA is a model which provides good forecasting on rainfall. ARMAX is another method which can compare two time series data. It can be used to compare the rainfall data with crop production, weather data and other causes related to rainfall.

Bibliography

Bez Abih, "Farmers' Response to Rainfall Variability and Crop Portfolio Choice." *Environment for Developement* (University of gothernburg), (2011): 1–28.

D. P. Acharjya, P. Kauser Ahmed, "A Survey on Big Data Analytics: Challenges, Open Research Issues and Tools." *International of Computer Science and Applications* 7, no. 2 (2016): 511–518.

Sotiris Archontoulis, Fernando E. Miguez. "Non linear Regression Models and Applications in Agricultural Research." *Statistical Concepts*. Agron, (2014): 786–798.

G. Arvind, P. Ashok Kumar, S. Girish Karthi *et al.* "Statistical Analysis of 30 Years Rainfall Data: A Case Study." *IOP Publishing* (2017): 1–10.

Woldeamlak Bewket. "Rainfall Variability and Crop Production in Ethiopia Case Study in the Amhara Region." *Conference Proceedings*, (2009): 823–836.

N. L. Chourasiya, P. Modi, N. Shaikh *et al.* "Crop Prediction Using Machine Learning." *IOSR Journal of Engineering (IOSR JEN)* (2019): 6–10.

EMC. *Data Science and Big Data Analytics*. Wiley, Indianapolis, IN, 2015.

Arnav Garg, Himanshu Pandey. "Rainfall Prediction Using Machine Learning." *International Journal of Innovative Science and Research Technology* 4, no. 5 (May 2019): 56–58.

GDP. "GPD in agriculture." https://tradingeconomics.com/india/gdp-from-agriculture, 2019.

Daume, hal. *A Course in Machine Learning*, ciml.info, 2012.

Abdul-Aziz Ibn Musah, Jianguo Du, Thomas Udimal *et al.* "The Nexus of Weather Extremes to Agriculture Production Indexes and the Future Risk in Ghana." *MDPI* 6, no. 4 (October 2018): 1–24.

Varun Khatri. "Application of Fuzzy Logic in Water Irrigation System." *International Research Journal of Engineering and Technology* 5 (April 2018): 3372–3375.

Mohan Krishna, Hari Krishna, Ashok Chakarvarthy. "A Fuzzy Environment Strategy For Optimal Agricultural Land Allocation In Krishna Delta." *International Research Journal of Computer Science* 5, no. 2 (February 2018): 57–64.

Amit Kumar, Manish Shrimali, Sukanya Saxena *et al.* "Forecasting Using Machine Learning." *International Journal of Recent Technology and Engineering* 7, no. 6C (April 2019): 38–41.

Arun Kumar, Alka Choudhry. *Descriptive Statistics*. Meerut: Krishna Prakasan Media Ltd., 2010.

Arun Kumar, Naveen Kumar, Vishal Vats. "Efficient Crop Yield Prediction Using Machine Learning Algorithms." *International Research Journal of Engineering and Technology* 5, no. 6 (June 2018): 3151–3159.

Rohitashw Kumar, Harender Raj Gautam. "Climate Change and Its Impact on Agricultural Productivity in India." *Journal of Climatology & Weather Forecasting* 2, no. 1 (April 2014): 1–3.

Conrad Kyei-Mensah, Rosina Kyerematen, Samuel Adu-Acheampong. "Impact of Rainfall Variability on Crop Production within the Worobong Ecological Area of Fanteakwa District, Ghana." Edited by Christos Tsadilas. *Advances in Agriculture*, May 2019: 1–8.

Konstantinos G. Liakos, Patrizia Busato, Dimitrios Moshou *et al.* "Machine Learning in Agriculture: A Review." *MDPI* (August 2018): 1–29.

David B. Lobella, Marshall B. Burke. "On the Use of Statistical Models to Predict Crop Yield Responses to Climate Change." *Agricultural and Forest Meteorology (Elsevier)* 150, no. 11, July 2010: 1443–1452.

K. Menaka, N. Yuvaraj. "A Survey on Crop Yield Prediction Models." *Indian Journal of Innovations and Developments* 5, no. 12 (October 2016): 1–7.

Francis Ndamani, Tsunemi Watanabe. "Rainfall Variability and Crop Production in Northern Ghana: The Case of Lawra District." *Conference Proceedings*, 2013: 1–9.

Kodimalar Palanivel, Chellammal Surianarayanan. "An Approach for Prediction of Crop Yield Using Machine Learning and Big Data Techniques." *International Journal of Computer Engineering and Technology* 10, no. 3 (June 2019): 110–118.

Aakash Parmar, Kinjal Mistree, Mithila Sompura. "Machine Learning Techniques for Rainfall Prediction: A Review." *International Conference on Innovations in information Embedded and Communication Systems*, 2017: 1–7.

Kenneth Peprah. "Rainfall and Temperature Correlation with Crop Yield: The Case of Asunafo Forest, Ghana." *International Journal of Science and Research* 3, no. 5 (May 2014): 784–789.

S. Poudel, Rajib Shaw. "The Relationships between Climate Variability and Crop Yield in Mountainous Environment." *MDPI* (March 2016): 1–19.

Kyada Pradip, Pravendra Kumar, Sojitra Manoj. "Rainfall Forecasting Using Artificial Neural Network (ANN) and Adaptive Neuro-Fuzzy Inference System (ANFIS) Models." *International Journal of Agriculture Sciences* 10 (2018): 6153–6159.

P. Priya, U. Muthaiah, M. Balamurugan. "Predicting Yield Of The Crop Using Machine Learning Algorithms." *International Journal Of Engineering Sciences & Research Technology* 7, no. 4 (April 2018): 1–7.

J. Refonaa, M. Lakshmi, Raza Abbas, Mohammad Raziullha. "Rainfall Prediction using Regression Model." *International Journal of Recent Technology and Engineering* 8, no. 2S3 (2019): 543–546.

Sujata Terdal Renuka. "Evaluation of Machine Learning Algorithms for Crop Yield Prediction." *International Journal of Engineering and Advanced Technology* 8, no. 6 (August 2019): 4082–4086.

V. Sellam, E. Poovammal. "Prediction of Crop Yield Using Regression Analysis." *Indian Journal of Science and Technology* 9, no. 38 (October 2016): 1–5.

Wenjiao Shi, Fulu Tao, Zhaol Zhang. "A Review on Statistical Models for Identifying Climate Contributions to Crop Yields." *Journal of Geographical Sciences (Science Press-Springer)* 23, no. 3 (2013): 567–576.

Kishor Shinde, Parag Khadke. "The Study of Influence of Rainfall on Crop Production in Maharashtra State of India." *Conference Proceedings*, (July 2017): 1–5.

Monira Sumi, Faisal Zaman, Hideo Hirose. "A Rainfall Forecasting Method Using Machine Learning Models And Its Application To The Fukuoka City Case." *International Journal Applied Mathematics and Computer Science* 22, no. 4 (2012): 841–854.

Kalpesh S. Tailor. "Statistical Analysis for Sugar cane Crop Production, Area under Cultivation and Total expenses for Olpad Region." *International Journal of Engineering and Management Research* 7, no. 1 (February 2017): 45–48.

G. Bala Sai Tarun, J. V. Sriram, K. Sairam *et al.* "Rainfall Prediction Using Machine Learning Techniques." *International Journal of Innovative Technology and Exploring Engineering* 8, no. 7 (May 2019): 957–963.

Christos Tzimopoulos, Christos Evangelides, Christos Vrekos *et al.* "Fuzzy Linear Regression of Rainfall-Atitude Relationship." *MDPI*, (August 2018): 1–9.

K. S. Varsha, Maya L. Pai. "Rainfall Prediction Using Fuzzy C-mean Clustering and Fuzzy Rule-Based Classification." *International Journal of Pure and Applied Mathematics* 119, no. 10 (2018): 597–605.

S. Veenadhari, Bharat Misra, C. D. Singh. "Machine Learning Approach for Forecasting Crop Yield Based on Climatic Parameters." International Conference on Computer Communication and Informatics, (January 2014): 1–6.

Lotfie A. Yousif, Abdelrahman A. Khatir, Faisal M. El-Hag. "Rainfall Variability and Its Implications for Agricultural Production in Gedarif State, Eastern Sudan." *African Journal of Agricultural Research* 13 (August 2018): 1577–1590.

Yousuf Zaman. "Machine Learning Model on Rainfall – A Predicted Approach for Bangladesh." Thesis, Dhaka, 2018.

D. S. Zingade, Omkar Buchade, Nilesh Mehta *et al.* "Crop Prediction System using Machine Learning." International Journal of Advance Engineering and Research Development 4, no. 5 (December 2017): 1–6.

Chapter 3

Discrimination between weed and crop via image analysis using machine learning algorithm

P. Amsini and R. Uma Rani

SRI SARADA COLLEGE FOR WOMEN (AUTONOMOUS)

3.1 Introduction

The agriculture field is working to save our environment by making certain exceptional changes in its practices to secure crop production. This approach is totally based on technology which can support soil preparation, planting and weed eliminating process. The most vital problems surface due to weeds which increase the biological competition with existing crop such as excess use of fertilizer, water and manual working hours. Manual work of weed inspection and removal takes a longer time. However, robotic applications are able to perform weed detection in excellent way. It is one kind of mechanical elimination task. Weed area classification using images taken in outdoor environment is very complicated due to random and uncontrolled light conditions. But images are evaluated through image processing techniques such as shape, size and texture features which make the classification of weed area easy. A strategy to employ cameras is required to determine the positions of weed in crops so that stereoscopic images can be processed. The image pixels are processed through various image processing algorithms which helps to differentiate between crops and weeds. Once the weed position accurately mapped, it is removed by robotic arms. This kind of automatic weed detection robotics work is based on sensors and image processing algorithm (Figure 3.1).

3.1.1 Problem of the statement

During the past years, weed detection was done physically by humans. Afterward with the innovation in technology, herbicides came into use to expel the weeds. Then image processing came into use for weeding. In this chapter, detection of weeds in the crop using image processing will be focused on. Manual weed sampling is time and cost intensive and therefore cannot be economical in a wider practice. Advanced fuzzy image processing based clustering is applied to segregate the weeds in crops to reduce the manual work. Image analytics is done by machine learning to accurately detect weed-infected areas.

3.1.2 Robust weed detection in image processing

The major benefits of automatic based weed systems which help to reduce the labor price and the usage of pesticides. Also, these systems locate the weeds in crops in an

Figure 3.1 Various kinds of weeds with wide leaves in crop images.

efficient way. In a survey done in Australia, it is reported that farmers spend about 1.5 billion dollars every 12 months for weed control activities. This is a robust method to increase the productivity and limit losses.

3.1.3 Objectives of proposed work

Machine learning algorithms are used for plant differentiation and weed detection with accuracy. These algorithms are used in real-time applications of nondestructive analysis of image objects.

3.2 Literature survey

An exhaustive research was done on several papers describing various methods adopted for weed detection. These papers were summarized as follows.

Image processing techniques and machine vision are broadly used in various fields such as agriculture industry or for detection of an object. The images are mathematically represented as rows and columns with red. So, automatic weed management systems are impacting the economy of the country green and blue channels. The weed-infected crop images are inspected either manually or automatically with robotics. Sometimes the weed and crops are of same color, in this condition we use the fuzzy image processing for oscillation basis. The absence of weeds in crops is detected using the advanced fuzzy set algorithms. The starting endeavors to distinguish weed seedlings using machine vision were centered on geometrical estimations such as shape, angle proportion, length region etc. Afterwards, color-based pictures were successfully used to identify weed infected area. The weed scope and weed patchiness are based on the computerized images where employing fuzzy-based calculations is used to select the site of the weed area. Tellaeche *et al.* (2007) proposed to apply k means and Bayesian-based algorithms for decision-making regarding spraying of adequate amount of herbicides. Fast Fourier transform process was applied for weed detection in various corn fields (Nejati *et al.*, 2008). Agriculture industry plays one of the most essential roles in economy of any country. These days, weeds are controlled by automatic robotic cultivators. The robotics captures the images of the crops and defines the weed-infected region. Then it eliminates the weed by spraying herbicide accurately on the weed. Fast Fourier transform algorithm is used in weeding robot to detect the accurate area of the weed. The process includes preprocessing, frequency and density filtering, and finally post-processing. In preprocessing, separation of background is done through Euclidean distance based on green and blue pixel channels. Density and frequency features are collected from preprocessed images and clustered by optimization classifiers. Finally, the weed area foreground and background are distinguished accurately.

Bhongale and Gore (2017) proposed a dilation and erosion method applied to segmentation process, it is used to classify the weeds in crops. In this method, a threshold value is applied to segment the weed area. These kinds of algorithms are used in agricultural and crop scouting. In crop scouting, applications are used for pest and weed detection in various plants which reduces the manual working time. The image processing algorithms are used to improve the accuracy of weed area detection in broad and narrow leaves. Yield is lost due to weeds in the crop, so their detection by automatic process enhances the yield. Dyrmann *et al.* (2017) proposed to detect the weeds using fully convolution neural network. GoogleNet is one of the networks which can process thousands of images to detect even the overlapping weeds. The network produces the predictive convergence map and bounding boxes. The convergence maps show if the weeds are present in crops. Output convergence maps are used to locate the weeds in the image. The proposed work helped to overcome the problem of detection of small weeds.

Pulido Rojas *et al.* (2019) proposed an application named Auto weed used in Australian rangelands. This real-time application detects the weed area by collecting the dataset of hyper-spectral images and thus classifies the weed area. Such robust methods help in improving the accuracy of weed area detection. Lot of research work is going on in the robotic weed control field. Robotic weed control contains four processes: mapping, control, guidance and detection of weeds. Weed control robotics are developed based on spectrum, image and spectral image-based methods to detect the weeds using aerial and ground photography. The image features are taken in numerical format to check whether it is a crop or weed, this functionality is known as image analytics. The most important part is to detect the features and classify them according to various features like shape, texture, statistical features, etc. Weeds reduce the crop production, so various applications are implemented which work better and reduce the human working time. The weed locations are identified through robotics, so a continuous research is going on in this area. The machine learning algorithms are applied to detect the weed infected region and measure the quantity of herbicides required. These automatic methods can surely help to enhance the economic level of a country by increasing the agricultural production. If any weeds are not detected by four-wheel robots, they move forward and check other fields. Color-based image processing is suggested by most of the authors for weed detection because it can capture variations of pixels in red, green and blue ranges. Sometimes oscillation occurs in weed area classification, but this problem is solved by advanced fuzzy set theories. From this literature survey, various weed processing algorithms are evaluated. Linear searching algorithm is used to detect the weed rows in crop farming fields.

3.3 Methodology

Weeds are detected using various machine learning techniques based on image processing methods. The weed detection in agricultural field uses various feature properties such as shape, size, texture features and spectral reflectance. The flow of proposed

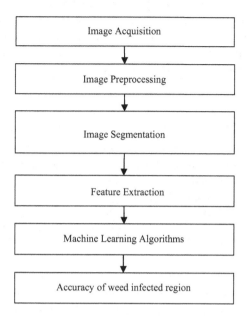

Figure 3.2 Flow of the proposed work.

work consists of image acquisition, preprocessing, segmentation and feature extraction as shown in Figure 3.2.

3.3.1 Image acquisition

The dataset comprises 60 images and it is accessible online. These pictures are taken from real-time robot bonirob. The images are of carrot plants for detecting inter- and intra-weeds (Haug and Ostermann, 2014). All these images are taken for the processing and provide better results for detection of the areas of weeds and plants.

3.3.2 Image preprocessing

An image Z with M rows and N columns and its intensity level range is measured as a collection of fuzzy singletons. The intensity range is in between L, 0 to 1. The membership value μ_{ik} with color intensity x_n x_m is considered. The contrast intensification operator was introduced by Zadeh in 1973. This operator is fully dependent on the membership function and applied for enhancement proposes. The constant operator is selected from image entropy value (e). So,

$$\text{InT} = \sum_{v=0}^{L-1} [\mu_{ik}' - e]^2 . J(k) \tag{3.1}$$

$$\sum_{v=0}^{L-1} J(k) = 1 \tag{3.2}$$

where, J(k) represents the frequency of intensity values. The spatial plane is changed by applying the intensification operators. Before these existing methods were proposed, various authors suggested filtering operations such as median, wiener and mode and then contrast level improvement was done through point operations. In previous works, e value was calculated through constant value 0.5. The "index of fuzziness" increases and decreases based upon the image contrast.

3.3.3 Image segmentation

Advanced fuzzy set theory is mostly used in real-time applications such as medical, satellite and agricultural field (Rani and Amsini 2019). In 1975, Zadeh pioneered another advanced fuzzy set called type II fuzzy set. Obviously, membership functions were defined by an expert based on his or her knowledge. These fuzzy set theories were applied to images that were poorly illuminated. The objects were barely visible in those images and furthermore uncertainties occurred. Type II fuzzy sets are the fuzzy sets for which the membership function is not a solitary value, meaning that every constituent is an interval.

A type II fuzzy set may be written as $A_{type\,2} = \{x, \mu_A(x) \mid x \varepsilon X\}$, where $\mu_A(x)$ is the type II membership function. An interval-based type II fuzzy set is distinct with its upper and lower membership values and practically represented as $\mu^{upper} = [\mu(x)]^{\alpha}$; $\mu^{lower} = [\mu(x)]^{1/\alpha}$. The type II fuzzy set considers the fuzzy membership function as fuzzy. The uncertainty is corresponding to upper and lower levels of the membership function. A lot of authors suggested that fuzzy set theory is superior in obtaining better results with precise values for better-quality analysis. Fuzzy set theory helps to improve the accuracy and reduces error due to oscillation. Much research is going on advanced fuzzy set theories such as mathematical modeling and robotics.

3.3.4 Proposed interval type II intuitionistic fuzzy c means with spatial triangular fuzzy number

This algorithm decreases various uncertainties in histopathology images. It comprises the following steps:

Step 1: Set the initial values for the centroids γ_i wherever i =1 to c.
Step 2: The primary memberships of higher $\bar{\mu}_{ik}$ and lesser membership of μ_{ik} are set according to two fuzzifiers m_1 and m_2 . The constant m satisfies the condition $m_1, m_2 \geq 1$. The two fuzzifiers help to construct footprint of uncertainty

$$\begin{cases} J_{m1}(U,v) = \sum_{j=1}^{N}\sum_{i=1}^{C} \left(\mu_{jk} \right)^{m1} d_{ik}^2 \\ \\ J_{m2}(U,V) = \sum_{j=1}^{N}\sum_{i=1}^{C} \left(\mu_{jk} \right)^{m2} d_{ik}^2 \end{cases} \qquad (3.3)$$

Step 3: The k represents the Euclidian distance d_{ik}^2 between pixel samples; the rate of spatial order is computed for every SP_{ij} pixel as follows:

$$\overline{\text{SPA}}_{ik} = \frac{\sum_{j=1}^{N} \overline{\mu_{ij}} \left(d_{kj}\right)^{-1}}{\sum_{j=1}^{N} \left(d_{kj}\right)^{-1}} \tag{3.4}$$

$$\underline{\text{SPA}}_{ik} = \frac{\sum_{j=1}^{N} \underline{\mu_{ij}} \left(d_{kj}\right)^{-1}}{\sum_{j=1}^{N} \left(d_{kj}\right)^{-1}} \tag{3.5}$$

The value of the spatial information is defuzzified as

$$\text{SPA}_{ik} = (\overline{\text{SPA}_{ik}} + \underline{\text{SPA}}_{ik}) / 2 \tag{3.6}$$

Step 4: Compute the matrix of membership grades U_{ik} by means of the distance calculation as

$$\text{NewDist}_{jk} = \left\| x_k - \gamma_i \right\|^2 \cdot \left(1 - \alpha e^{-\text{SP}_{ik}}\right) \tag{3.7}$$

Step 5: Update the centroid of clusters as

$$\text{DD}^j = \left[\gamma_1^j, \gamma_2^j, \ldots, \gamma_c^j\right] \tag{3.8}$$

Step 6: Lastly, verify the stop condition, if max $(|J^{(j+1)} - J^{(j)}|)$, go to subsequent process otherwise go to step 2.

This proposed algorithm detects the weed from various crop images by selecting GLCM features from segmented image. Segmentation of the weed from crops and soil in the input image properly. These results are shown in Figure 3.3. The weed and crops in same color and size, so machine learning techniques are applied to detect and separate weeds and plant areas. The results guide the automatic sprayer and weed removal robotics. This proposed fuzzy c means algorithm improves the results and accurately detects weed areas. This process helps to reduce the manual work and improve farming process.

3.4 Feature extraction

The structure represents texture-based local properties of micro-texture and macro textures represent the spatial texture of narrow properties. These properties are not similar between the image pixels. The statistical features based method builds relationship among the gray levels. One-pixel based classifiers known as first order derivative and more than two pixels based classifier is known as second order derivative. The second order derivative is one of the greatest derivate in texture analysis. In biomedical and remote sensing fields these kinds of textures are evaluated.

The feature of the first order histogram provides different statistical properties such as four statistical moments of the intensity histogram of an image. These depend only on

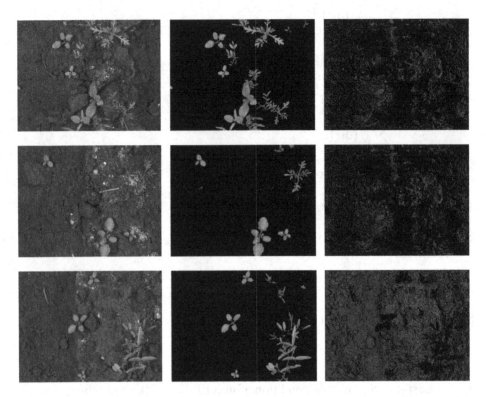

Figure 3.3 Weed detection using type II spatial intuitionistic fuzzy c means with tri-angular fuzzy number.

entity pixel values and not on the interaction or else co-occurrence of neighboring pixel values. The first order histogram statistics are mean, skewness, kurtosis and entropy. The GLCM is used to extract the second order texture in sequence from the images.

3.4.1 Mean

The mean describes the center value of the intensity pixels and it is denoted by the image features:

$$\text{Mean} = \sum_{b=0}^{1-1} p(s).$$

$$(3.9)$$

3.4.2 Skewness

Skewness (sw) is a computation of the asymmetry of the probability distribution of a real-valued random variable. The skewness value is negative or positive. The probability is considered as total number of neighborhood (p(s)) centered pixels by number of gray level pixels(s):

$$sw = \frac{1}{\sigma_s^3} \sum_{b=0}^{l-1} (s - \bar{s})^3 p(s). \tag{3.10}$$

3.4.3 Kurtosis

The kurtosis is the measurement of "peakedness" of the probability distribution of real value-based random variable:

$$kw = \frac{1}{\sigma_s^4} \sum_{b=0}^{l-1} (s - \bar{s})^4 p(s) - 3. \tag{3.11}$$

3.4.4 Entropy (En)

The concept of entropy comes from thermodynamics. Entropy is used to calculate the randomness and used to describe the texture of the input image. This value is between maximum of all elements of co-occurrence matrix. The pixels of i and j character-ize the coefficient of concurrence matrix J(i,j) and S represented as dimension of co-occurrence matrix. It is defined as

$$En = \sum_{i=0}^{S-1} \sum_{j=0}^{s-1} J(i,j)(-\ln(J(i,j)). \tag{3.12}$$

3.4.5 Contrast (Cn)

The contrast is a measure consisting of pixel intensity and its neighbor over the image and it belongs to the brightness of the object:

$$Cn = \sum_{i=0}^{S-1} \sum_{j=0}^{s-1} (i-j)^2 J(i,j). \tag{3.13}$$

3.4.6 Energy (Eg)

Energy is defined as extension of pixel pair replication and calculates the uniformity:

$$Eg = \sqrt{\sum_{i=0}^{S-1} \sum_{j=0}^{s-1} J^2(i,j)}. \tag{3.14}$$

After segmentation, all gray level co-occurrence matrix features are extracted from the segmented images and the weed and plant area are distinguished using machine learning techniques. The neural network and support vector machine (SVM) classifier help to detect weed infected area with accuracy. The gray level co-occurrence matrix feature values are given in Table 3.1.

Table 3.1 GLCM features extracted from image dataset

Segmented image	Mean	Skewness	Kurtosis	Entropy	Contrast	Energy
Image 1	0.2224	11.1851	1.0081e+004	6.6791	0.3425	0.2987
Image 2	0.1592	8.9473	2.0918e+004	7.6791	0.9402	0.1645
Image 3	0.2088	15.9818	1.5981e+004	9.6791	0.3501	0.2077

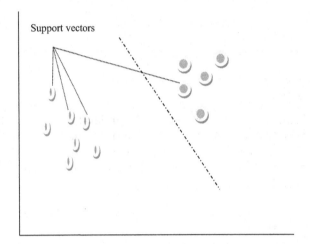

Figure 3.4 SVMs with linear hyperplane.

About 60 image features are taken for SVM and neural network classifiers to detect the weed and plants dissimilarities. Both classifiers are applied to detect the weed and plant regions via bounding boxes as shown in Figure 3.4. These kinds of machine learning algorithms are used in robotics for selecting the region and spraying herbicides automatically.

3.5 Support vector machine classifier

The machine learning algorithm of SVM is supervised method and used for both regression and classification problems. In SVM algorithm, given plot of image data with n number of features each feature belongs to a particular dimension. Here the two classes are distinguished as weed and as plant. The SVM is a linear hyper plane between the two classes as shown in Figure 3.4.

3.6 Neural network classifier

The image data are trained through the convolution neural network. The input is given as the feature-based image which separates the weed and plants in the segmented image. The convolution operation is performed on the matrix of pixels and trained

Table 3.2 Accuracy of weed infected region measured by SVM and neural network classifiers

Serial number	Accuracy of weed infected area by neural network (%)	Accuracy of weed infected area by SVM (%)
Image 1	89	94
Image 2	90	92
Image 3	91	91

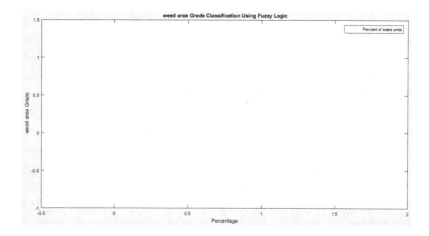

Figure 3.5 Weed area grade measured by fuzzy logic: Image 1.

through extracted features. The two layers involved are fully connected and pooling layers. The pooling layer helps to oversimplify the output of convolution layers. Both classifiers work well and extract weed infected region accurately, details are given in Table 3.2.

The weed affected region in plant is identified through fuzzy logic as shown in Figures 3.5–3.7.

The results of differentiation between weed and plants are shown in Figure 3.8. The bounding box represents disparity between plants and weeds. The differentiation between weeds and plants is done using machine learning algorithms. Such automatic detection methods are used in smart agricultural fields.

3.7 Conclusion

Agriculture plays a significant role in improving the economic condition of any country. The crop growth is reduced due to weeds. Earlier the weeds were detected manually, however it is a very expensive and time consuming process. Currently, weed detection is done by robotics, automatic sprayer and weed cutting are thus used.

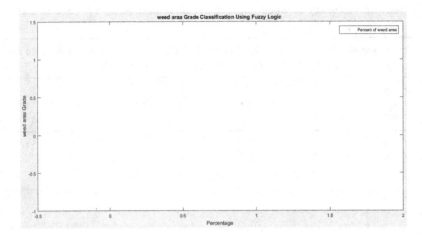

Figure 3.6 Weed area grade measured by fuzzy logic: Image 2.

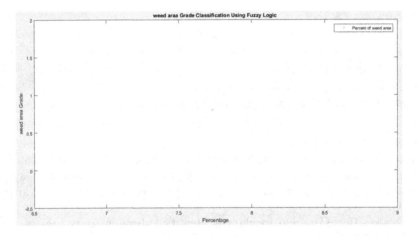

Figure 3.7 Weed area grade measured by fuzzy logic: Image 3.

This kind of robotics works based upon sensors, image processing and machine learning algorithms were embedded within hardware devices. This chapter proposes image processing and machine learning algorithms used to identify weeds from plants. The proposed algorithm classifies the weed and plants from the soil and then GLCM features are selected from segmented images. These features are applied in SVM and neural network machine learning algorithms which helps to differentiate the weed from plants accurately using bounding boxes. This automatic algorithm helps the smart agricultural field to detect the weed area in its early stage automatically and thus reduces the manual work and excess use of herbicides.

Figure 3.8 Image analytics through machine learning techniques.

References

Kalyani Bhongale, Sonal Gore, *Weed recognition system for crops in farms using image processing techniques and smart herbicide sprayer robot*, IJCESR, 4(7) (2017), 2393–8374.

Xavier P. Burgos-Artizzu, Angela Ribeiro, Maria Guijarro, Gonzalo Pajares, *Real-time image processing for crop/weed discrimination in maize fields. Computers and Electronics in Agriculture*, 75(2) (2010), 337–346.

Fenil Dankhara, Kartik Patel, Nishant Doshi, *Analysis of robust weed detection techniques based on the Internet of Things (IoT), Procedia Computer Science,* 160 (2019), 696–701.

M. Dyrmann, R. N. Jørgensen, H. S. Midtiby, *Robo Weed Support – Detection of weed locations in leaf occluded cereal crops using a fully convolutional neural network*, Precision Agriculture (ECPA), 8(2) (2017), 842–847.

Sebastian Haug, Jörn Ostermann, "*A Crop/Weed Field Image Dataset for the Evaluation of Computer Vision Based Precision Agriculture Tasks,*" 2014.

G. Jones, E. Ch. Gée, E. F. Truchetet, *Modeling Agronomic Images for Weed Detection and Comparison of Crop/Weed Discrimination Algorithm Performance.* Springer, Berlin, 2008.

Hossein Nejati, Zohreh Azimifar, Mohsenn Zamani, *Using Fast Fourier Transform for Weed Detection in Corn Fields.* IEEE, New York, 2008.

Camilo A. Pulido Rojas, Leonardo E. Solaque Guzman, Nelson F. Velasco Toledo, *Deep Weeds: A multiclass weed species image dataset for deep learning, Scientific Reports*, 9 (2019), 2058.

Grianggai Samseemoung, Peeyush Soni, Hemantha P. W. Jayasuriya, Vilas M. Salokhe, *Application of low altitude remote sensing (LARS) platform for monitoring crop growth and weed infestation in a soya bean plantation, Precision Agriculture*, 13(6) (2012): 611–627.

Alberto Tellaeche, Xavier P. Burgos Artizzu, Gonzalo Pajares, Angela Ribeiro, "*A Vision-Based Hybrid Classifier for Weeds Detection in Precision Agriculture Through the Bayesian and Fuzzy k-Means Paradigms,*" In: E. Corchado, J. M. Corchado, A. Abraham (eds), *Innovations in Hybrid Intelligent Systems*, Springer, Berlin, Heidelberg, 2007, pp. 72–79.

Hongkun Tian, Tianhai Wang, Yadong Liu, Xi Qiao, Yanzhou Li, *Computer vision technology in agricultural automation—A review, Information Processing in Agriculture*, 7 (2019), 2214–3173.

R. Uma Rani, P. Amsini, *Detection of weed area in crop images using type II intuitionistic fuzzy set algorithm, JETIR*, 6(3) (2019), 262–266.

Aichen Wanga, Wen Zhang, Xinhua Weia, *A review on weed detection using ground-based machine vision and image processing techniques, Computers and Electronics in Agriculture*, 158 (2019), 226–240.

Chapter 4

Bio-inspired optimization algorithms for machine learning in agriculture applications

P.R. MahiDar

SASI INSTITUTE OF TECHNOLOGY AND ENGINEERING

Deepika Ghai

LOVELY PROFESSIONAL UNIVERSITY

4.1 Introduction

With the increase in population, the food production also needs to be increased drastically with the limited available sources. To enhance productivity in farming, so many sophisticated tools are there. Nowadays, internet of things (IoT) and machine learning (ML) are playing a big role in agriculture industry [1, 2].

4.1.1 Machine learning in agriculture

ML techniques along with image processing algorithm are used in precision agriculture to increase food production in agricultural fields. Previous data available at different stages of farming are used to predict the conditions to improve production by means of ML algorithms. ML algorithms are useful at different stages of agriculture, such as yield prediction, disease detection, soil management, water management, weed detection, crop and seed quality and livestock production, etc., to predict conditions for improvement of crop.

In order to increase productivity, yield prediction, one of the main topics in precision farming, is of great importance. Yield prediction is important for growth, crop supply matching demand and crop management. Such forecasts help farmers prevent market demand and supply imbalances triggered or accelerated by the crop quality. Disease detection and pest control both are important for improvement of productivity and crop quality. Pesticide spraying over the cropping area is an effective method for disease control. ML algorithms are useful for the automatic detection of infected plants and it is less time consuming compared with normal eye examination.

Weed detection and control is a big issue in cultivation. Precise weed identification is important for better agriculture, because it is difficult to detect and distinguish weeds from crops. In water management, evapotranspiration data are analyzed using ML algorithms for water resource management and crop production. ML algorithms are useful to estimate soil conditions, temperature and moisture content. ML algorithms are also applied for seed and crop quality estimation. The livestock production refers to studies designed to accurately predict and estimate farming parameters in order to maximize the production system's economic performance.

4.2 Optimization in machine learning

ML facilitates systems to recognize patterns from current available algorithms and datasets and develop feasible solution concepts. In ML algorithms, to recognize the patterns it is required to feed the system with the required algorithms and huge data in advance. Then ML carries out some tasks. First it finds, extracts and summarizes the pertinent data. Then it makes predictions by analyzing the summarized data. After that, it determines probabilities for particular results. Then adapts to assured developments. Finally, optimization of these developments is done based on recognized patterns. The process of obtaining knowledge from available data is called learning process. In the approach of attempts to acquire information from available data, the results are probabilities rather than certainties. So, it is clear that optimization is part of ML. Nature gives a lot of motivation to derive solutions for complex and hard problems of optimization. However, it exhibits extremely dynamic, varied, complex, robust and fascinating occurrences. Nature always discovers the optimal solution, and to find the solutions it preserves perfect stability between its components. This becomes the inspiration for Bio-inspired optimization algorithms (BIOAs). A number of bio-inspired optimization methods have been recently designed to solve optimization problems. The next section presents different BIOAs [1, 2].

4.2.1 Bio-inspired optimization algorithms

There are two branches of optimization of problems known as exact methods and heuristics. BIOAs solve heuristic problems by imitating the strategies of nature. There are two main and booming classes in BIOAs. Those are evolutionary algorithms (EAs) and swarm-based algorithms (SBAs). EAs are influenced by the evolution in nature, and SBAs are driven by cooperative behavior of animals.

4.2.1.1 Evolutionary algorithms

Evolutionary computation is a concept in ML whose main objective is to increase the gain knowledge from the phenomena of collectiveness in adaptive population for problem-solvers utilizing the iterative progress including selection, growth, development, reproduction and survival as in population. EAs are algorithms of optimization which are nondeterministic or cost based. Genetic algorithm, genetic programming, evolutionary strategy, differential evolution (DE) and paddy field algorithm come under EAs. They are stochastic search algorithms based on the population which performs in the best-to-survive approach. Every method originates by producing a primary population with feasible number of solutions and progresses repetitively from one generation to the next generation to find optimum solution. In the algorithm, consecutive iterations are considered for fitness. This is called fitness choice. This happens in a decision population. For next generation, best solutions will survive [3,4].

4.2.1.1.1 GENETIC ALGORITHMS

In 1975, Holland proposed a genetic algorithm which is a stochastic algorithm based on evolution for optimization. This algorithm follows the theory of the "survival of the fittest" proposed by Charles Darwin. First, a solution population called chromosomes is initialized. It represents the problem in the bit vector form. Then the fitness of every chromosome is evaluated by use of a suitable problem's fitness function. The selection of best chromosomes depends on their fitness and the best selected chromosomes are sent into the mating pool, which go through crossover and mutation operations to give rise to a new set of improved solutions. Genetic algorithm gives good solution in complex search spaces where traditional methods fail. It has drawbacks that it is complicated to operate on dynamic datasets and not suitable for solving constraint problems of optimization [5,6].

4.2.1.1.2 GENETIC PROGRAMMING

In 1992, Koza proposed genetic programming. It is an expansion to genetic algorithm. Genetic programming characterizes a tree-type non-direct encoding of a probable solution, and computer program might be used in which search is applied directly to the solution. Genetic programming takes up variable-length representation, whereas fixed-length encoding is adopted by genetic algorithm. In genetic programming, population generates variation in values of genes and also in individual structures. Genetic programming involves four steps. They are generation of an initial population for computer programs. It executes each program, depending on how well it solves the problem and assigns fitness value to it. Then, using best currently available programs it creates new computer programs by mutations and crossover [7,8].

4.2.1.1.3 EVOLUTION STRATEGIES

This is developed through natural selection as inspired from the theory of adaptation and evolution. It deals with micro- or genomic-level data. It is a global optimization algorithm for regulating the distribution of mutations, it utilizes self-adaptive mechanisms. These self-adaptive mechanisms involve search progress optimization by developing solutions for the problem and also a few parameters for transforming the solutions. Evolution strategies use different selection and sampling schemes such as $(1+1)$, $(\mu+\lambda)$ and (μ, λ) evaluation strategies. In $(1+1)$ strategy, one of its parent's real-evaluated vectors of object variables is formed by applying mutation to each object variable with an equal standard deviation. The resulting individual is then compared and contrasted with its parent and the better one succeeds to the next generation as a parent and the remaining solutions are left. In $(\mu+\lambda)$ strategy, by selecting μ parents from the current generation, λ children are generated. These λ children are generated through some mutation operators on selected parents. From $(\mu+\lambda)$ population, that is, μ parents and λ children, only the best μ succeeds to the next generation. In (μ, λ) evaluation strategy, λ children (with $\lambda \geq \mu$) are generated by selecting current generation of μ parents and only the best μ children are selected to the further generation, and parents are discarded completely [9,10].

4.2.1.1.4 DIFFERENTIAL EVOLUTION

Storn and Price in 1995 projected differential evolution (DE) algorithm, which belongs to EAs [11]. It is similar to genetic algorithm as for optimal solution searches the individual populations are used. In DE, arithmetic combinations of individuals are called mutation, whereas in genetic algorithm small modifications to an individual's genes are called mutations. Hence, DE mutation is not dependent upon a predefined probability density function. DE implementation is simple and shows fast convergence. Greedy nature causes noise effect on the performance of DE [12].

4.2.1.1.5 PADDY FIELD ALGORITHM

Premaratne *et al.* [13] proposed paddy field algorithm in 2009. It functions on principle of reproduction depending on closeness to the population density and global solution. It is analogous to population of plants. It uses pollination and dispersal strategy. Paddy field algorithm comprises five basic steps called sowing, selection, seeding, pollination and dispersion. This algorithm starts by scattering seeds randomly in an irregular field (sowing). Next selects best solutions and dig over worst solutions (selection). During seeding, the seeds drop into proffered places and best plants (taller) continue to grow and produce huge number of seeds. For seed propagation, animals or wind pollination is used. When carried by the wind, the probability of pollination for pollen is increased due to high population density. Every plant's seeds are distributed to avoid the pollen from getting stuck at local minima [14,15].

4.2.1.2 Swarm intelligence

This optimization algorithm is motivated by organism's collective social behavior. This involves the implementation as a collective intelligence problem-solving method of simple agent groups focused on the real-world insect swarm actions. The word "swarm" is used to describe the particles "uneven movements in problem space" [16].

By considering five fundamental principles, swarm intelligence can be described as follows:

i.   Proximity principle: Simple space and time calculations can be carried out by population.
ii. Quality principle: In environment, the population must be able to react to quality factors.
iii. Diverse response principle: The activities along extremely narrow channels should not be committed by the population.
iv. Stability principle: Every time with respect to changes to environment, the population should not change their behavior.
v.   Adaptability principle: If the computational price is worth it, then it can alter the mode of behavior of the population.

4.2.1.2.1 PARTICLE SWARM OPTIMIZATION

In the year 1995, Kennedy and Eberhart proposed the particle swarm optimization (PSO) technique. This algorithm is motivated by the common behavior of bird groups for food searching. In PSO, the members without mass and without volume depending on the velocities and accelerations in the direction of the best mode of behavior are called particles. Each particle within the swarm represents a solution in the specified high-dimensional space. It considers four vectors which are its current best positions at the moment. Depending on its neighborhood and its velocity, the best position is found, then based on the best position its position is altered in the search space achieved by itself (p_{best}) and on the best position achieved by its neighborhood (g_{best}) during the search process. PSO can be described by a sequence of steps. First, allocate a swarming random location in the searching space and compute the fitness function of each particle. Individual fitness value of particle is compared with the best fitness value, and if the current fitness value is better than the best fitness value, then this value is set as the best solution (x_i as p_i) and the current particle location. The particle is identified with the best fitness value and the velocities and locations of the particles are changed using the same procedure repeated until final requirements are met. PSO is easier to implement, efficient in maintaining the diversity and it has a more memory capability than genetic algorithms [17–20].

4.2.1.2.2 ANT COLONY OPTIMIZATION

Dorigo and Di Caro predicted this algorithm in 1999, which is one of the most popular SBAs. It is a meta-heuristic algorithm that is inspired by ants' forestry behavior, called stigmergy. It enables indirect contact between self-organizing growing systems by moving individuals across their local environment. It depends on the collaborative actions of group of ants and their shortest path finding capability to find food sources from their nest and then by tracing out pheromone trails. After that, ants choose the pathway in which a decision is based on probability influenced by the quantity of pheromone: the stronger pheromone trace, the greater its desirability. This behavior leads to a self-mechanism which leads to the formation of pathways marked by high pheromone concentration, as ants in turn drop pheromone in the direction they follow. By modeling and simulating ant's behavior, techniques such as brood sorting, nest building are developed. This can be built for complex combinatorial problems of optimization. This algorithm was developed in 1996 and named as ant system to solve travelling salesman problem. Ant's colony optimization algorithm is implemented with three functional blocks: ant solutions construct, pheromone update and daemon actions [21, 22].

4.2.1.2.3 ARTIFICIAL BEE COLONY ALGORITHM

The foraging behavior and mating behaviors of bees are the motivation for artificial bee colony algorithm. A bee chooses a food source by waiting for a decision in dance area and is called onlooker. Some bees visit the food source before calling employed bees. Random search is performed by scout bees to find new sources of food. The food source location and the amount of nectar correspond to optimization problem's possible solution and the quality (fitness) of the solution. In this two-dimensional search space, a cluster of virtual bees is created which begins to move about randomly. When some target nectar is found by the bees, they interact. The intensity of bee interactions is used to obtain the solution of the problem. This algorithm runs in three phases. First is search process, the next phase is reproduction and the final phase is replacement of bee and selection [23,24].

4.2.1.2.4 FISH SWARM ALGORITHM

In 2002, fish swarm algorithm was proposed by Li *et al.*, which is a new swarm intelligent evolutionary computation method based on population. Natural schooling behavior of fish is the motivation for this algorithm. This algorithm achieves global optimization by avoiding local minimums. The fish swarm algorithm follows three types of behaviors denoted as finding food, swarming to a danger and following to increase probability of achieving an unbeaten result. These steps can be applied to an optimization problem [25,26].

4.2.1.2.5 INTELLIGENT WATER DROPS ALGORITHM

Hamed Shah Hosseini proposed intelligent water drops (IWD) algorithm in 2007. It is a revolutionary approach focused on population. It is enthused by the natural river system processes which constitute the actions that occur between flow of river water and the environmental changes in which the river flows. It has two important properties:

i.   River carried soil
ii. River's current velocity of flow

The environment of the river water flow is analogous to the optimization problem. The IWD move from their present position to their next place in steps of a discrete finite length. The soil's dissimilarity between two locations defines the velocity. The velocity of the IWD is improved by an amount inversely proportional to soil added to the IWD. The time taken to pass the water drop from one location to the other location is inversely (and nonlinearly) proportional to the quantity of mud added to the IWD. The time taken and distance between the two locations is inversely proportional to each other. The time taken and velocity of the IWD are directly proportional to each other. An IWD chooses the low soil paths to higher soil paths on their strata. To introduce path selection behavior, a uniform random distribution among the soils of the available paths is used, so that the probability of next path is inversely proportional to the available path soils. The lower the path soil, the greater the chance it has of being selected by the IWD [27–29].

4.2.1.2.6 BACTERIAL FORAGING OPTIMIZATION ALGORITHM

The bacterial foraging optimization algorithm was proposed in 2002 by Passino. This algorithm comprises mainly three mechanisms called chemo taxis, reproduction and elimination–dispersal. Bacteria get together in the nutrient-rich areas in an unstructured manner called chemo taxis. In reproduction, superlative personalized bacteria survive and spread their genetic characteristics to next population. Some part of bacteria become moderate and scatter into arbitrary positions of environment to avoid getting trapped in local optima is called elimination dispersal [30].

4.3 Conclusion

This chapter gives a brief introduction of ML in agriculture applications and BIOA applicable in ML. Classification plays an important role in ML and deep learning.

While working with huge data, there is a need of optimization. This chapter motivated and gave guidelines to the readers to apply BIOAs in ML to solve various optimization problems.

References

1. S. Binitha, S.S. Sathya, "A Survey of Bio-Inspired Optimization Algorithms", *International Journal of Soft Computing and Engineering*, vol. 2, no. 2, pp. 137–151, 2012.
2. J. Qiu, Q. Wu, G. Ding, Y. Xu, S. Feng, "A Survey of Machine Learning for Big Data Processing", *EURASIP Journal on Advances in Signal Processing*, vol. 67, no. 1, pp. 1–16, 2016.
3. D. Soni, "Introduction to Evolutionary Algorithms Optimization by Natural Selection", Towards Data Science, 2018 https://towardsdatascience.com/introduction-to-evolutionary-algorithms-a8594b484ac
4. A.N. Sloss, S. Gustafson, "2019 Evolutionary Algorithms Review", pp. 1–37, 2019, arXiv:1906.08870.
5. Genetic Algorithms, pp. 53–57, https://shodhganga.inflibnet.ac.in/bitstream/10603/35732/10/10_chapter%202.pdf
6. Y.H. Liao, C.T. Sun, "An Educational Genetic Algorithms Learning Tool", *IEEE Transactions on Education*, vol. 44, no. 2, pp. 210, 2001.
7. L. Vanneschi, R. Poli, "Genetic Programming - Introduction, Applications, Theory and Open Issues", In: Rozenberg G., Back T., Kok J.N. (eds) *Handbook of Natural Computing*, Springer, Berlin, Heidelberg, pp. 709–739, 2012. https://doi.org/10.1007/978-3-540-92910-9_24.
8. A. Pétrowski, S.B. Hamida, "Genetic Programming for Machine Learning", *Evolutionary Algorithms*, pp. 183–216, 2017 https://doi.org/10.1002/9781119136378.
9. P. Oliveira, F. Portela, M.F. Santos, A. Abelha, J. Machado, "Machine-Learning an Overview of Optimization Techniques", *Recent Advances in Computer Science*, Bentham Science Publishers, United Arab Emirates, pp. 52–56.
10. A.H. Gandomi, A.H. Alavi, "Krill Herd Algorithm: A New Bio-Inspired Optimization Algorithm", *Communications in Nonlinear Science and Numerical Simulation*, vol. 17, no. 12, pp. 4831–4845, 2012.
11. R. Storn, K. Price, "Differential Evolution - A Simple and Efficient Adaptive Scheme for Global Optimization Over Continuous Spaces", *Journal of Global Optimization*, vol. 11, no. 4, pp. 341–359, 1997.
12. S. Das, P.N. Suganthan, "Differential Evolution: A Survey of the State-of-the-Art", *IEEE Transactions on Evolutionary Computation*, vol. 15, no. 1, pp. 1–42, 2011.
13. U. Premaratne, J. Samarabandu, T. Sidhu, "A New Biologically Inspired Optimization Algorithm," in: *4th International Conference on Industrial and Information Systems, Sri Lanka*, December 28–31, 2009, pp. 279–284.
14. X. Kong, Y.L. Chen, W. Xie, X. Wu, "A Novel Paddy Field Algorithm based on Pattern Search Method", in: *2012 IEEE International Conference on Information and Automation, Shenyang, China*, June 6–8, 2012, pp. 686–690.
15. K.G. Liakos, P. Busato, D. Moshou, S. Pearson, D. Bochtis, "Machine Learning in Agriculture: A Review", *Sensors*, vol. 18, pp. 1–29, 2018.
16. H.R. Ahmed, J.L. Glasgow, "Swarm Intelligence: Concepts, Models and Applications", Technical Report 2012-585, School of Computing Queen's University Kingston, Ontario, Canada K7L3N6, February 2012.
17. Z. Zhang, W. Huangfu, K. Long, X. Zhang, X. Liu, B. Zhong, "On the Designing Principles and Optimization Approaches of Bio-Inspired Self-Organized Network: A Survey", *Science China Information Sciences*, vol. 56, no. 7, pp. 1–28, 2013.

18. G.B. Huang, X. Ding, H. Zhou, "Optimization Method based Extreme Learning Machine for Classification", *Neurocomputing*, vol. 74, no. 1–3, pp. 155–163, 2010.
19. J. Kennedy, R. Eberhart, "Particle Swarm Optimization", in: *Proceedings of the 1995 IEEE International Conference on Neural Networks, Perth, WA, Australia*, November 1995, pp. 1942–1948.
20. D. Bratton, J. Kennedy, "Defining a Standard for Particle Swarm Optimization", in: *2007 IEEE Swarm Intelligence Symposium, Honolulu, HI, USA*, April 1–5, 2007, pp. 120–127.
21. M. Dorigo, M. Birattari, T. Stützle, "Ant Colony Optimization", *IEEE Computational Intelligence Magazine*, vol. 1, no. 4, pp. 28–39, December 2006.
22. Y. Pei, W. Wang, S. Zhang, "Basic Ant Colony Optimization", in: *International Conference on Computer Science and Electronics Engineering*, Hangzhou, China, March 23–25, 2012, pp. 665–667.
23. Y. Xu, P. Fan, Ling Yuan, "A Simple and Efficient Artificial Bee Colony Algorithm", *Mathematical Problems in Engineering*, vol. 2013, pp. 1–9, 2012.
24. A. Aderhold, K. Diwold, A. Scheidler, M. Middendorf, "Artificial Bee Colony Optimization: A New Selection Scheme and Its Performance", *Nature Inspired Cooperative Strategies for Optimization*, vol. 284, pp. 283–294, 2010.
25. M. Neshat, A. Adeli, G. Sepidnam, M. Sargolzaei, A.N. Toosi, "A Review of Artificial Fish Swarm Optimization Methods and Applications", *International Journal on Smart Sensing and Intelligent Systems*, vol. 5, no. 1, pp. 107–148, March 2012.
26. F.S. Lobato, V. Steffen Jr., "Fish Swarm Optimization Algorithm Applied to Engineering System Design", *Latin American Journal of Solids and Structures*, vol.11, no.1, pp. 143–156, 2014.
27. H.S. Hossein, "Intelligent Water Drops Algorithm A New Optimization Method for Solving the Multiple Knapsack Problem", *International Journal of Intelligent Computing and Cybernetics*, vol. 1, pp. 193–212, 2008.
28. H.S. Hossein, "An Approach to Continuous Optimization by the Intelligent Water Drops Algorithm", *Procedia - Social and Behavioral Sciences*, vol, 32, pp. 224–229, 2012.
29. X.S. Yang, *Nature-Inspired Optimization Algorithms*, School of Science and Technology, Middlesex University London, Elsevier, 2014.
30. S. Das, A. Biswas, S. Dasgupta, A. Abraham, "Bacterial Foraging Optimization Algorithm: Theoretical Foundations, Analysis, and Applications", *Foundations of Computational Intelligence*, vol. 3, pp. 23–55, 2009.

Chapter 5

Agricultural modernization with forecasting stages and machine learning

A.K. Awasthi and Arun Kumar Garov

LOVELY PROFESSIONAL UNIVERSITY

5.1 Introduction

People have consistently been extraordinary anticipators, due to the necessity of bread and butter from the beginning. Humans have predominant physical qualities, and they are keen observers. In any situation, we all homo sapiens are the extraordinary organizers, a feature that has given us points of interest for the field of forecasting.

For various reasons there is always a need for making predictions and forecasting is very much required in daily life. Forecasting may be future planning in the form of 1 day, 1 month or 1 year and so on based on the present time and situation.

Forecasting is done according to different sectors and different situations, for example, in the agriculture sector while planting the seeds for definite crops in particular season, the farmer forecasts for better harvesting results.

Another example can be taken from the energy world where forecasting is used to decide the number of power generation plants to be set up as required in the future. Similarly, many other examples can be cited to show the necessity of forecasting.

Let us have a look at different fields or sectors where we are using forecasting.

5.2 Where can we use forecasting?

Forecasting can be done in many areas for future requirements. But the question is if forecasting is not able to change the present, then will it be of use in future. Forecasting is used in different areas for different situations, some areas where forecasting is done are given in Figure 5.1.

Thus, it is no big surprise that we have been endeavoring to sharpen our skills of forecasting throughout the ages. Even though we are still a long way from arriving at the objective of perfection, it is astonishing how far we have come to give the best possible result, thanks to the previous researchers. Before going further, we should review the seven key periods of forecasting.

5.3 Stages of forecast

In previous paragraphs, an idea for the necessity of forecasting in different sectors was discussed. Now we should understand the different stages of predictive analysis and forecasting. These stages are represented in the flow chart (Figure 5.2).

According to the given stages, all stages can be linked with a primary era in the following way (Figure 5.3):

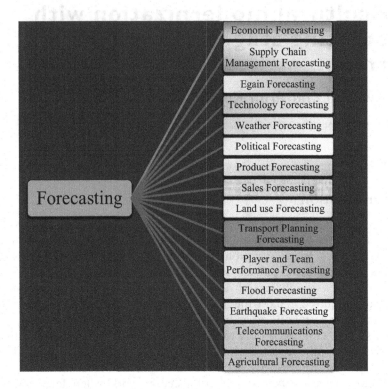

Figure 5.1 Classification of forecasting area.

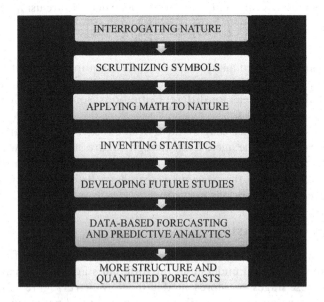

Figure 5.2 Stages of predictive analysis and forecasting.

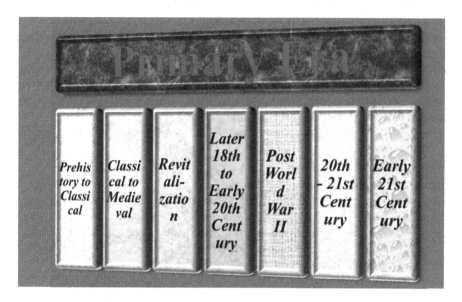

Figure 5.3 Classification of primary era.

The relation between types of primary era and stages of forecasting

Stage number	Stages	Primary era
I.	Interrogating nature	Prehistory to classical
2.	Scrutinizing symbols	Classical to medieval
3.	Applying math to nature	Revitalization
4.	Inventing statistics	Later 18th to early 20th century
5.	Developing future studies	Post-World War II
6.	Data-based forecasting and predictive analytics	20th–21st century
7.	Quantified forecasting and more structure	Early 21st century

5.3.1 Details of forecasting stages

5.3.1.1 Stage 1. Interrogating nature

Prehistoric to classical primary era

During prehistoric times, humans started hunting and looking around themselves. They thought about the requirement of the materials for the future and tried to fulfil these requirements.

Humans considered how the birds are flying. Early people followed the practices, for example, of divination of watching the travel of flying creatures to know what will happen. At an essential level, perhaps there were useful advantages of these efforts.

For instance, they noticed that when seagulls quit flying and take asylum someplace, it can imply that a tempest is drawing nearer.

Humans watched the natural phenomena and looked at the behavior of birds and animals. They started to experiment to fly like birds. So, they made structures like wings, and tried to fly. In Roman times, people rehearsed in the craft of prognostication of the universe of winged animals. There were different less exquisite divination practices also, for example, haruspicy, which was the investigation of the insides of sacrificed creatures to get hints of what will come.

5.3.1.2 Stage 2. Scrutinizing symbols

Classical to medieval primary era

The procedure called "sortilege" or "cleromancy" includes anticipating the future from sticks, beans or different things drawn indiscriminately from an assortment. Such practices appear since the beginning in a wide range of societies, from Judeo-Christian customs (throwing dice on different places in the Bible) to the Chinese convention of the I. Ching, which comes from bone divination.

These types of rules were expelled from the understanding of regular examples and we moved one step nearer to image-ruled determining. In any event, expecting that there's no fundamental legitimacy to cleromancy, it can at times give gainful impacts to chiefs. I. Ching, for instance, depends on both the flipping of coins and the consequent understanding of the content to which the coin hurls a point. As the content is translated, individuals can bring both their cognizant and oblivious personalities to a given issue. In specific situations, this may relax the limitations on the customary idea and lead to progressively creative arrangements and choices.

5.3.1.3 Stage 3. Applying math to nature

Revitalization primary era

Throughout long stretches of written history, people became cautious onlookers of numerous sorts of naturals designs, particularly the examples of the heavenly bodies. Through sheer observation, a few specialists had the option to foresee the movements of the stars and planets with incredible exactitude.

Nicolaus Copernicus (1473–1543) wrote the *Revolutions of the Heavenly Spheres*. He says that the humans change according to the universe. Through a powerful mix of rationale, arithmetic and experimental perception, he had the option to show that the earth revolves around the sun instead of the other way around. His thoughts were later affirmed by Galileo Galilei and others.

Through such means, people turned out to be more aware of what was going on in their close planetary system, and our prescient forces become significantly progressively noteworthy. In the long run, masters like Isaac Newton and Gottfried Leibniz created analytics. This was a tremendous advance forward in the order of anticipation. All things considered, math is eventually the investigation of how things change, giving a "structure for demonstrating frameworks in which there is change, and an approach to conclude the expectations of such models", as indicated by Prof. Daniel Kleitman of MIT.

5.3.1.4 Stage 4. Inventing statistics

Later 18th to early 20th century primary era

There are some factual strategies that are more than two centuries old, for example, the likelihood hypothesis started in the 17th century. Presently, we think that it came out in the late nineteenth to mid-twentieth century. Karl Pearson presented the item minute relationship coefficient, and John Galton created key ideas, for example, standard deviation, connection and even relapse investigation. The splendid Ronald Fisher built up the invalid theory and numerous other key ideas. Such thoughts stay essential to present-day information science, artificial intelligence (AI) and prescient examination.

It comes down to the fact that mankind at last had a method for gathering, investigating and displaying numerical information identified with probabilities. This insight enables scientists and examiners to quantify, convey and, some of the time, even control vulnerability.

5.3.1.5 Stage 5. Developing future studies (evolution of future study)

Post World War II primary era

The beginnings of future examinations lie with individuals, for example, Samuel Madden and H.G. Wells, who utilized a blend of creative mind and trend watching to make surmises about the development of society and, on account of Wells, its most significant innovations. Be that as it may, "future investigations" just turned into a conventional control after World War II, when advances and social frameworks were changing at such a quick pace that the future showed up to be less sure than in past periods.

The World Future Society was established in 1966, and the World Futures Studies Federation was established in 1973. During those years, the originally alleged futurists developed. Alvin Toffler, the creator of the compelling and prominent Future Shock, helped transform the point of things to come into a standard subject. During this period, futurists built up various philosophies, including "genuine games", situation arranging, model and recreation creation and visioning. In any case, futurists were (and are) regularly reprimanded for flawed gauging or by evading determining through and through for future creation or future expectations.

5.3.1.6 Stage 6. Data-based forecasting and predictive analytics

20th–21st century primary era

In present investigation as it goes back to previous years similar to current insights itself, yet it was not until the 1950s that different associations started utilizing PC (computer-based modeling) based on demonstrating everything from climate examples to credit dangers. During the 1970s, the celebrated Black Scholes model was created to anticipate the best costs for investment opportunities, and during the 1990s investigation turned out to be broadly utilized for everything from web searches to baseball line-ups. The firmly related field of AI started prospering during the 1990s. After that, prescient models became normal for all intents and purposes of each datum-based order, from science to showcasing to criminal equity.

5.3.1.7 Stage 7. Quantified forecasts and predictive analytics

Early 21st century primary era

Even though this technique additionally has profound recorded roots, as of late as inquired about led by Philip Tetlock and others illustrated—through estimating competitions held by Intelligence Advanced Research Projects Agency—that individuals can turn out to be observationally better at anticipating. The examination recommends that great forecasters can even show others how to gauge all the more viably. One key to accomplishing this is the measurement of anticipating by doling out probabilities.

5.4 Advantages of forecasting

Some significant points of forecasting are expressed below.

1. It empowers an organization to submit its possessions with the best faithfully to gain advantage over the long term.
2. It encourages the advancement of the new object, by distinguishing future-required designs.
3. Forecasting by advancing the investment of the whole association in this procedure gives chances of cooperation and thus realizes unity and coordination.
4. Forecasts, as observed by administrators, constrain thinking ahead, looking to the future and accommodating it.
5. Forecasting is a fundamental element of arrangement that supplies inevitable actualities and essential data.
6. Forecasting gives an approach to compelling coordination and control. It requires data about different outer and inner variables.
7. Forecast, as a methodical endeavor to test the future by induction from well-established actualities, incorporates all administration arrangements with the goal of bringing together divisional and departmental plans.
8. The vulnerability of future occasions can be distinguished and defeats be determined. Subsequently, it will prompt achievement in the organization.

5.5 Disadvantages of forecasting

The essential weakness of forecasting is equivalent to that of some other strategy for anticipating the future: No one can tell properly about the future. Any unexpected thought, however small, can waste the prediction of the future. Additionally, some forecasting strategies may utilize similar information to convey broadly various estimates. For example, one forecasting technique can show that loan costs will rise, while another will outline that rates will hold consistent or decay.

5.6 Impediments (limitations) of forecasting

5.6.1 Just estimates

No one knows about the future, one can just take an assumption of the future. The future is not fixed even if the best technique of forecasting is used. But forecasting is just an estimation of the future. No one can get 100% success in the prediction of future incidents. Therefore, even the best-laid plans may fail. This consistently remains one of the greatest limitations of the forecasting.

5.6.2 In view of assumptions

The premise of any anticipating technique are suspicions, approximations, ordinary conditions and so forth. This makes these conjectures untrustworthy. So, one should consistently remember the innate confinements of forecasting and be careful in being over-dependent on them.

5.6.3 Time and cost factors

The information and data required to make formal forecasts are commonly an important component. Furthermore, the assortment and classification of records need time and cost. The transformation of subjective information into quantitative information is additionally another component. One must be cautious that the time, cost and exertion spent in estimation must not exceed the genuine advantages from such figures.

5.7 Introduction to agriculture

In the world, the most important livelihood source is agriculture. Before the industrial revolution, agriculture was a major source of economy. With many trade options coming up, many are dependent for their income on agriculture. Agriculture is the most peaceful and friendly activity for environment, and it is a very reliable and honest source of human livelihood. In developing countries, many people rely on agriculture for their livelihood. Some people have jobs/businesses but they also use agriculture as a side business. It is not limited only to cultivation and farmers. Poultry, beekeeping, dairy, forestry, sericulture are also included in agriculture.

5.8 Classification for agriculture

One of the most world-prevalent activities is agriculture. But this is not similar in all countries. Cultivation of certain plants and raising of domesticated animals are done for the process of producing foods, fiber and some other desired products. The word "Farming" is used for the agricultural process. In the agricultural field, many changes in farming practices have come up in the 20th century especially due to the advent of machine learning technique used in agricultural forecasting. Some criteria and classification are adopted for agriculture as shown in Figure 5.4.

5.9 Types and branches of agriculture

Some general types of agriculture around the world are shown in Figure 5.5.

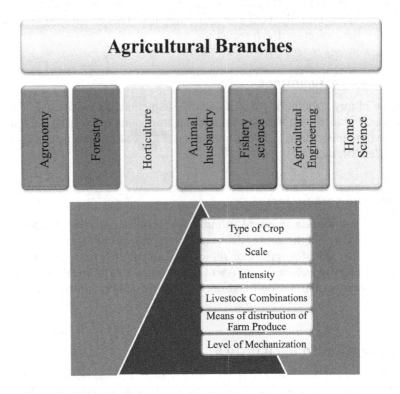

Figure 5.4 Classification of agriculture.

Figure 5.5 General types of agriculture.

Following are the branches of agriculture:

In agriculture, the most important is the yield of the crops. For a better yield of crops, farmers are using a yield prediction.

Yield prediction is the most important and popular topic in precision agriculture. It is represented by yield mapping and estimation of yield, using comparing the supply and demand of crop and crop management. According to the past data, we are going far away to simple predictions. Presently we are using various technologies to make most of the yield for farmers and population. These technologies are AI, machine learning, neural network, etc.

5.10 Fundamentals of farming seasons for better yields

A fundamental for seasons of farming is 1 year. Farming season has the following fundamentals for better yields of crop, likewise temperature and moisture as conditions should be suitable for better yield of the crop. Under farming season, appropriate climate change especially helps farmers for farming plants and grow crops.

Raining seasons depend on the temperature and potential of evapotranspiration. In the world, some areas are not better for the growth of the crops. According to the distance from Equator, different countries of the world have different seasons.

In India, agricultural crop season starts in July and ends in June. This season is divided into two parts. According to monsoon, first is kharif and the second is rabi. Kharif and rabi words are Arabic words, where the meaning of these words are autumn and spring, respectively. Kharif crop starts in July, and rabi crop grows during October–March, which is the winter season. Rabi crops are wheat, chickpea, oat, barley, peas, sunflower, etc., and kharif season crops are bajra, jowar, maize, soybean, etc.

Around the world, wheat grows in different planting and harvest seasons. In the season of the wheat crop, the market prices fluctuate according to the yield.

In the United States and China, there are two types of wheat crop: first is winter wheat and the second is spring wheat. China is one of the large producers of wheat compared with other countries. After China, the nest good producers of wheat are India, the United States, etc.

5.11 Knowledge of machine learning

Machine learning enables the system with the capability to automatically explore, enhance and improve according to different situations without being programmed. Machine learning is centered on the development of intelligent computer programs that can process the data and utilize it further.

Machine learning is powered by statistics, calculus, linear algebra and probability statistics. Calculus tells us how to learn and optimize our mode, linear algebra makes running these algorithms feasible on the massive datasets and the probability helps in predicting the likelihood of an event occurring (Figure 5.6).

We make a set of attributes defined by an individual example. Sets of the attribute are known as features and variables. Binary, numeric and ordinal can also represent these features. The performance metric is used for calculating the performance of machine learning.

In today's world, we are surrounded by lots of technologies. So, it is better to stay up to date with new emerging technologies.

5.12 Technology in agriculture

The agricultural industry has a vast role in technology. The innovation of technology makes agriculture modernized. Various machinery and tools have helped farmers to play a vital role in developing the economy of a country (Figures 5.7 and 5.8).

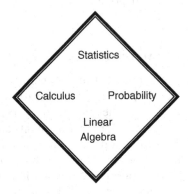

Figure 5.6 Component of machine learning.

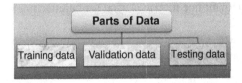

Figure 5.7 Technological innovation impacting agriculture.

5.13 Requirement for agriculture

5.13.1 Data

For any analysis, data are important. Data analysis, machine learning and AI involve one common and important feature called "data". If we do not have data, then we cannot train a model. Big enterprises spend money in big quantities to collect the data. Data can be in any form such as value, text, picture, etc..

We can divide the data into three parts for machine learning:

5.13.1.1 Training data

It is a part of the data used for training of the model. From these data, our model learns and looks at the input and output.

Figure 5.8 Technologies for agriculture development.

5.13.1.2 Validation data

This part of the data is used for continuous assessment of the model. Here training dataset fits along with the changes involving hyperparameters. When the model starts working, then it is used for training.

5.13.1.3 Testing data

For unbiased evaluation in testing, the data model should be fully trained. In any model testing, data will be input and the model will predict some values without referring the actual output value. Hence testing data are useful for prediction. The predicted value obtained from the model is compared with actual output value for analyzing the result. As in any experiment, this process helps us to know how much a model is perfect from training dataset at the time of training.

After learning the need and types of data, its processing is most important.

In this process, the concept of machine learning arises. So now, we must understand the development of machine learning.

5.14 Development of machine learning

Machine learning technology is growing day by day in different sectors for analysis and prediction with the help of training data. So, training data are a key factor for machine learning. It tells us about the use of AI, so it is also used in agriculture.

Before applying any data for prediction through machine learning, some basic factors are needed.

5.14.1 First factor

Machine learning needs knowledge about around-the-world activities and tasks for training computers. It can explore themselves to educate themselves.

5.14.2 Second factor

This factor is digital data or information collected and made accessible for the analytics process.

5.14.3 Third factor

The third factor is where digital changes are made available for all technology-based environments and devices.

See one example of the latest technology. Technologies and deep learning algorithms on a drone are used to collect the data of crops and soil monitored by a software. Fertility of the soil is controlled using the software. Some companies are developing robots and automation tools for agriculture field to form effective ways to save a crop and also to protect them from weeds. Agricultural spray machines are designed for spraying accurate weedicides on the plants and in accurate amount to reduce expenditures.

5.15 Development of different areas through machine learning

Machine learning is developing with technologies of big data and other fast computer devices. In the field of agriculture, machine learning is creating some new opportunities to understand the different types of data processes related to environmental functions. Machine learning can be converted to a scientific formulation which will give the capability to learn without programming of the device. It is used in different areas such as medicines, meteorology, biochemistry, bio-informatics, robotics, economic sciences, climatology and food security.

5.16 Machine learning methods

In machine learning agriculture, it learns through agricultural processes to derive methods. In machine learning we have those types of datasets which depend on examples. An individual example is also used in examples of datasets. Characteristics of these datasets are known as variables or helpers. These features can also be described as numerical, binary and features. The performance of machine learning is calculated from performance metrics.

The machine learning model obtains experience with time and then improves its performance. Some statistical and mathematical models are used to determine the performance of the machine learning model and machine learning algorithms. When the learning process is completed, then the model may be used to classify and make the assumption and to test data. It can be achieved after the training process completion.

Methods and applications of machine learning in different sectors are given below.

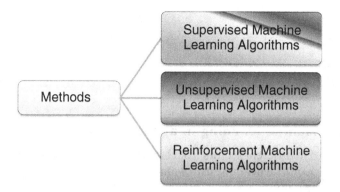

5.16.1 Supervised machine learning algorithm

Supervised learning is a method used to enable machines to classify/predict object problems or situations based on the data fed to the machine.

Example

Suppose we take data of circle, triangle and square labels in the labeled data. We have a training model and we know the answer. It is very important in supervised learning that you already know the answer about lots of the given information which is coming out. When we have a huge couple of data coming in and new data coming out, we can train the model. The model now knows the difference between circles, triangles, squares and other shapes that we trained it to identify. We can send the squares and circles to predict a top on the square and second on the circle. This functionality is used in agriculture. In the field of agriculture, there is huge data for prediction with some assumptions.

5.16.2 Unsupervised machine learning algorithm

Unsupervised learning is a machine learning model that finds the hint in the unlabeled data.

So, in the previous example to identify what the circle is, what the triangle is, and what the square is, it looks at the dimensions of the figure or preferably it looks together at the number of corners. Several models have three corners, two corners, one corner or no corners and the model labels a filter to true altogether.

5.16.3 Reinforcement machine learning algorithm

Here, the agent learns the property of behaviors to the environment by the performance of the act and checks the results of action.

Application of machine learning in different sectors:

1. Healthcare
2. Sentiment analysis

3. Fraud detection
4. E-commerce
5. Oil and gas
6. Transportation
7. Marketing and sales
8. Agricultural forecasting

5.17 Uses of machine learning in agriculture

AI technique is used in different sectors from home to offices, and presently in agriculture also. In the agricultural field, the use of machine learning increases the productivity and quality of the crop.

5.17.1 Retailers

The seed retailers use this agriculture technology to churn the data to create better crops. While the pest control companies are using it to identify various bacteria, bugs and vermins.

5.17.2 Increase the yield of crops by AI

AI technologies are used in agriculture to determine which crops will give better yield under which conditions. And it also determines which weather conditions will provide the highest return of yield.

5.17.3 Bug identification and solution

AI is also used to kill bugs and vermin. Some apps are used to find bugs. Here pictures of bugs are taken by an app known as PestID. When the app finds the bug, it provides immediate solutions and helps to further actions. It provides a list of the chemicals to kill the bugs.

5.18 A century of crop protection

Since this chapter is related to forecasting, we must have an idea of the development of agricultural stages so as to link these stages with machine learning. During the past 100 years, there was an indifference in the technology and most of the development is seen in the last 5 years (Figure 5.9).

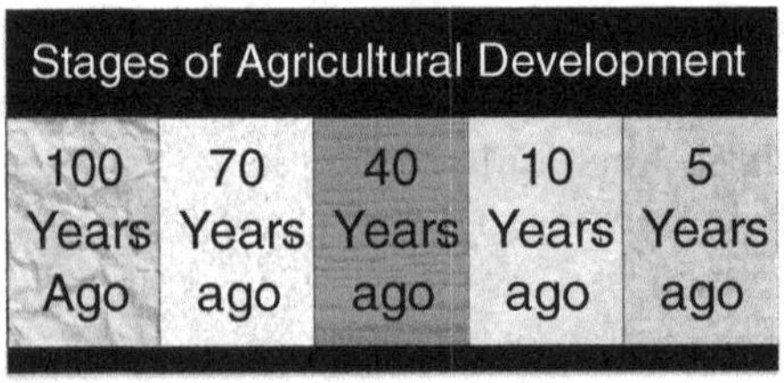

Stages of Agricultural Development				
100 Years Ago	70 Years ago	40 Years ago	10 Years ago	5 Years ago

Figure 5.9 Stages of agricultural development.

5.18.1 Hundred years ago

About 100 years ago, agriculture came into existence as a livelihood. At the same time, farmers grew some crops for the livelihood of their families.

Also, in this period farming was astonishingly difficult as compared to the present. At that time farmers physically afforded to grow crops for a long time. It is clear that the crop farming process using plough was harder and time consuming, so it was difficult in itself. In farming by plough lot of animal power and manpower was required. It means farmers were growing few crops and caring for livestock with their families. They were also facing the problem of poverty as well as plant diseases.

5.18.2 Seventy years ago

Before hybrid seeds became widely available, seeds bred for disease and/or pest resistance were rare, if available at all. In this period, some primary tools such as hand weeding, tillage, and crop rotation were used.

5.18.3 Forty years ago

Labor-intensive methods such as hand weeding and tillage were used by farmers in the field, without using chemicals.

5.18.4 Ten years ago

As farmers faced problems of insects, they started using chemical insecticides.

5.18.5 Five years ago

Farmers did not have data analytics and accurate equipment. At that time farmers were not capable of protecting the crop because they used to investigate the production by eye to take any action. In these activities, farmers were spending a lot of time.

Presently, farmers have many options to make a better choice. There are various options available today, such as digital tools, historical data, precision technologies and different types of crop protection methods. These tools help to grow the crop effectively with less effect on the environment.

5.19 Case study

We know that agriculture is an important part of gross domestic product. This project cussed in the advantage of insurance companies so they have efficient insurance coverage. In this project, we have taken two test datasets. One is 2CSV files and another is image dataset.

CSV file has lots of features like temperature, humanity, pressure and precipitation type, such as snow, rain and all other weather conditions and also wind speed visibility. Here we are predicting a crop field condition. The second dataset is the image dataset which consists of 10 images of crop fields. With drone images, we have manually counted the number of crops in each image. Here we have used Convolutional Neural Network (CNN) regression to predict the contradicting crops (Figures 5.10 and 5.11).

CountyNar	State	Latitude	Longitude	Date	apparentT	apparentT	cloudCove	dewPoint	humidity	precipInte	precipInte	precipProb	precipAcc	precipType	precipType	precipType	pressure	temperatu	temperatu	visibility	windBeari	windSpeed	NDVI	Day
Adams	Washingtc	46.92984	-118.352	########	18.61	-3.01	0	6.77	0.69	0	0	0	0	0	0	0	1027.95	23.93	6.96	10	9	3.8	136.1797	
Adams	Washingtc	47.15033	-118.959	########	19.67	-0.74	0	6.66	0.65	0	0	0	0	0	0	0	1028.26	25.88	8.71	10	352	6.03	135.6975	
Adams	Washingtc	46.81169	-118.695	########	20.66	-0.14	0	6.55	0.67	0	0	0	0	0	0	0	1028.29	24.67	8.26	10	25	3.59	135.677	
Adams	Washingtc	47.16234	-118.7	########	19.69	-2.66	0.03	7.32	0.69	0	0	0	0	0	0	0	1027.74	25.48	8.1	10	1	5.18	135.0058	
Adams	Washingtc	47.15751	-118.434	########	18.82	-3.04	0.04	7.62	0.7	0	0	0	0	0	0	0	1027.16	24.83	8.32	9.99	5	4.69	134.8039	
Adams	Washingtc	47.00689	-118.51	########	19.24	-2.58	0.04	7.45	0.7	0	0	0	0	0	0	0	1027.68	25.35	7.98	10	3	4.86	134.6575	
Alfalfa	Oklahoma	36.50822	-98.4543	########	56.86	10.06	0.04	21.74	0.51	0	0	0	0	0	0	0	1015.05	56.86	24.22	9.95	352	9.71	150.4279	
Alfalfa	Oklahoma	36.6507	-98.4447	########	56.32	9.57	0.04	20.59	0.49	0	0	0	0	0	0	0	1015.14	56.32	23.91	9.93	354	10.05	150.2179	
Alfalfa	Oklahoma	36.70245	-98.3199	########	56.89	9.68	0.05	20.8	0.5	0	0	0	0	0	0	0	1015.13	56.89	24.06	9.93	352	10.01	150.1431	
Alfalfa	Oklahoma	36.62878	-98.5398	########	56.15	9.32	0.04	20.1	0.49	0	0	0	0	0	0	0	1015.16	56.15	23.79	9.92	354	10.13	149.3406	
Alfalfa	Oklahoma	36.98881	-98.4046	########	52.51	8.05	0.04	18.12	0.5	0	0	0	0	0	0	0	1015.43	52.51	22.64	9.93	0	11.89	148.1095	
Alfalfa	Oklahoma	36.99583	-98.3309	########	52.53	8.14	0.04	18.33	0.5	0	0	0	0	0	0	0	1015.41	52.53	22.69	9.94	0	11.86	148.0441	
Allen	Kansas	37.82964	-95.5129	########	58.43	12.48	0.08	32.08	0.7	0	0	0	0	0	0	0	1014.64	58.43	25.28	9.23	347	5.7	147.5343	
Allen	Kansas	37.93584	-95.297	########	59.09	13.32	0.1	33.33	0.71	0	0	0	0	0	0	0	1014.73	59.09	25.75	9.24	344	5.21	146.8112	
Allen	Kansas	37.7755	-95.5177	########	59.72	13.36	0.08	33.35	0.71	0	0	0	0	0	0	0	1014.36	59.72	25.9	9.22	344	5.04	146.53	
Allen	Kansas	37.8864	-95.403	########	58.49	12.72	0.08	32.27	0.7	0	0	0	0	0	0	0	1014.68	58.49	25.5	9.26	346	5.53	145.9264	
Allen	Kansas	37.9449	-95.1587	########	59.34	13.76	0.09	33.89	0.71	0	0	0	0	0	0	0	1014.74	59.34	26.09	9.27	341	4.91	145.4953	
Allen	Kansas	37.82396	-95.4433	########	59.89	13.79	0.08	33.79	0.7	0	0	0	0	0	0	0	1014.34	59.89	26.28	9.26	342	4.86	144.9338	
Anderson	Kansas	38.07577	-95.1568	########	59.59	13.67	0.09	34.23	0.72	0	0	0	0	0	0	0	1014.71	59.59	26.04	9.23	338	5.83	146.8117	
Anderson	Kansas	38.09794	-95.4483	########	57.21	11.22	0.09	30.4	0.7	0	0	0	0	0	0	0	1015.02	57.21	24.28	9.19	347	6.74	146.4638	
Anderson	Kansas	38.15425	-95.0905	########	58.79	12.89	0.08	32.29	0.72	0	0	0	0	0	0	0	1014.99	58.79	25	9.1	339	7.51	145.7837	
Anderson	Kansas	38.20991	-95.2243	########	53.78	7.99	0.06	25.25	0.69	0	0	0	0	0	0	0	1015.5	53.78	21.91	8.9	351	10.46	143.9019	
Anderson	Kansas	38.34689	-95.1191	########	54.04	8.31	0.05	24.62	0.68	0	0	0	0	0	0	0	1015.67	54.04	22.1	8.88	349	11.04	143.5907	
Anderson	Kansas	38.23728	-95.3387	########	54.67	8.05	0.06	24.91	0.67	0	0	0	0	0	0	0	1015.5	54.67	21.87	8.99	352	10.25	143.0336	
Archer	Texas	33.40353	-98.5972	########	78.75	32.37	0	50.37	0.63	0	0	0	0	0	0	0	1012.43	78.75	40.68	9.35	203	7.91	150.9683	
Archer	Texas	33.41	-98.9153	########	77.99	28.38	0	47.08	0.61	0	0	0	0	0	0	0	1012.06	77.99	38.04	9.16	207	6.88	146.0862	
Archer	Texas	33.74978	-98.841	########	77.55	23.83	0	42.95	0.58	0	0	0	0	0	0	0	1012.08	77.55	34.78	10	215	4.64	145.2708	
Archer	Texas	33.59957	-98.4543	########	77.89	28.13	0	46.98	0.62	0	0	0	0	0	0	0	1012.11	77.89	37.64	10	207	6.14	144.9792	
Archer	Texas	33.48708	-98.9502	########	77.86	27.78	0	46.56	0.6	0	0	0	0	0	0	0	1012.06	77.86	37.62	9.21	208	6.69	144.6725	
Archer	Texas	33.79193	-98.4531	########	76.68	22.86	0	43.3	0.6	0	0	0	0	0	0	0	1012.35	76.68	34.15	10	212	4.25	142.6651	
Armstrong	Texas	35.05299	-101.295	########	70.07	7.04	0	16.43	0.3	0	0	0	0	0	0	0	1011.37	70.07	22.14	10	305	4.49	148.5028	
Armstrong	Texas	35.17797	-101.491	########	70.74	6.91	0	16.31	0.3	0	0	0	0	0	0	0	1011.31	70.74	22.13	10	303	4.55	148.4128	

Figure 5.10 CSV file having test data.

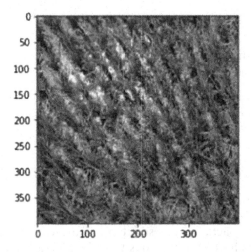

Figure 5.11 Crop development obtained by testing data.

Next, we discuss the preprocessing of the dataset and different machine learning algorithms used in this case study. So, first in the preprocessing part, we have to input all the set variable data using one odd in coding, then we have to normalize the data using max scalar. Then, we have to fill nine values by the mean values of that respective columns. Afterward we split whole dataset in the two parts, the first part is the training set, the second is testing set. Training set consists of 70% of the data and the second part consists 30%. Now we come to the different algorithms used. First, we used random forest, after that we applied support vector regression then, deep neural networks for regression.

To explain in detail the case study we studied the crops in the field that can be counted using the images. We used ten images taken by a drone and counted them manually with several crops per image. It took about 2 hours to count this. We then took eight of the images for training a model and two images for testing a model to see error rates. We used CNN model and modified it for predictive purposes by including a dancing layer with a linear activation function in the output layer. We normalized the images to 400×400 pixels with the length. So, feeding images into the model became easier. A model has about 14 hidden layers with a 25% dropout function. Since we do not have a large dataset, we cannot expect very large accuracy but we can predict a moderate level of accuracy based on the dataset (Figures 5.12–5.14).

5.20 Result and conclusion

Here we have used three methods:

1. Random forecast
2. Support vector regression
3. Deep neural network

```
[ ]   1 import keras
      2 from keras.models import Sequential
      3 from keras.layers import Dense, Dropout, Flatten
      4 from keras.layers import Conv2D, MaxPooling2D
      5 from keras.utils import to_categorical
      6 from keras.preprocessing import image
      7 import numpy as np
      8 import pandas as pd
      9 import matplotlib.pyplot as plt
     10 from sklearn.model_selection import train_test_split
     11 from tqdm import tqdm
     12 %matplotlib inline

[ ]   1 train = pd.read_csv('/content/drive/My Drive/Bennet/number_final.csv')    # reading the csv file
      2 train.head()        # printing first five rows of the file
```

Figure 5.12 Model programming graph.

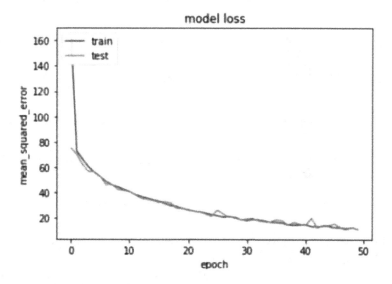

Figure 5.13 Model test and training graph.

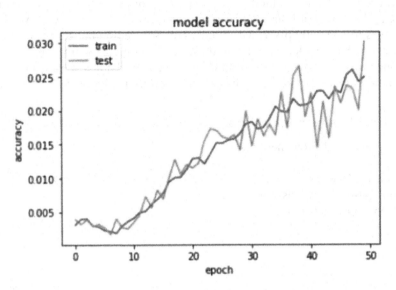

Figure 5.14 Model accuracy graph.

We have used the parameters of mean absolute error (MAE) and mean square error (MSE) and compared different models. We can see from the given bar graph that a good result is shown by deep neural network and also the MAE and MSE have lowest value in the represented in model.

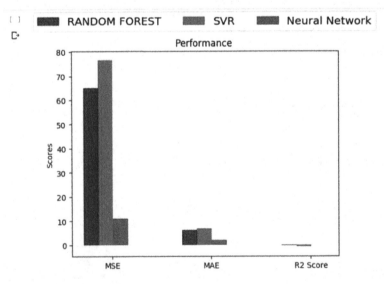

Now we will see the graph of the neural network. In this graph, we can see directly that the MSE and MAE are decreasing and the accuracy is increasing with the increasing number of crops. Now we will see the CNN case which is used for counting the images of crops (Figure 5.15).

```
[ ]   1 plt.plot(history.history['mean_absolute_error'])
      2 plt.plot(history.history['val_mean_absolute_error'])
      3 plt.title('model mean_absolute_error')
      4 plt.ylabel('mean_absolute_error')
      5 plt.xlabel('epoch')
      6 plt.legend(['train', 'test'], loc='upper left')
      7 plt.show()
      8 | # summarize history for loss
      9 plt.plot(history.history['loss'])
     10 plt.plot(history.history['val_loss'])
     11 plt.title('model mean squared error')
     12 plt.ylabel('mean_squared_error')
     13 plt.xlabel('epoch')
     14 plt.legend(['train', 'test'], loc='upper left')
     15 plt.show()
```

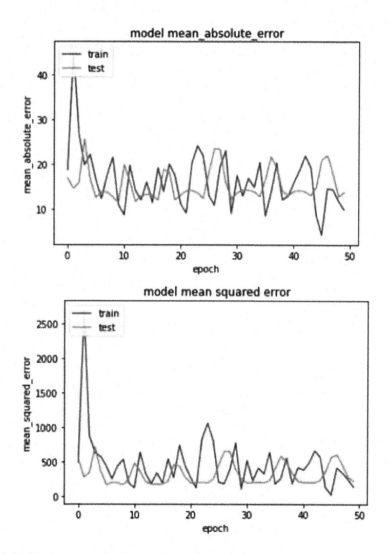

Figure 5.15 Model mean absolute error and squared error graph.

In CNN, you can see that with the increasing number of crops MSE is decreasing when the results between test data and train data were compared. According to the performance of model, we can say that the model is good. So, we can conclude that deep neural network process performs best for accuracy of crop yield prediction problem.

Bibliography

Awasthi, A. K., Garov, A. K., & Koul, S. (2020). Decision Making Model for Stock Index. *Journal of Xidian University*, 14, pp. 1261–1265.

Awasthi, A. K., Garov, A. K., & Kumar S. (2018). Integrating MATLAB and Python. *JETIR*, 5, pp. 91–99.

Ord, J. K., Fildes, R., & Kourentzes, N. (2017). *Principles of Business Forecasting* (2nd ed.). Wessex Press Publishing Co.

Snehal, S. D. (2014). Agricultural Crop Yield Prediction Using Artificial. *International Journal of Innovative Research in Electrical, Electronic*, 1 (1), pp. 683–686.

Chapter 6

Classification of segmented image using increased global contrast for paddy plant disease

Md Abdul Muqueem, G. Raju and Govind Singh Patel

LOVELY PROFESSIONAL UNIVERSITY

Seema Nayak

IIMT COLLEGE OF ENGINEERING

6.1 Introduction

An image conveys information. It consists of prime elements called pixels, where each pixel is represented by some value, with the function of two coordinates X and Y in mathematical representation. An image consists of organized pixels in rows and columns.

6.1.1 Greyscale image

The greyscale images consist of 256 grey tones of colors. The main feature of the greyscale image is the equal distribution of colors red, green and blue. It carries only intensity information of individual pixels as a single sample and is composed of a wide range of shades. It is also called monochrome image. The contrast ranges from black to white, black is taken as weakest intensity and white as the strongest intensity value. It is an 8-bit image which constitutes 256 shades.

6.1.2 Color image

A color image is a two-dimensional image usually represented by three different colors per pixel. It constitutes 24 bits per image. These 24 bits are divided into 3 equal parts, each of 8 bits which give information of intensity values of red, green and blue with 8 bits of red, 8 bits of green and 8 bits of blue information. The mixing of these colors gives good appearance to the image.

6.1.3 Binary image

Binary image pixel contains two grey level values. Grey level "0" constitutes black and grey level "1" constitutes white color. The pixel value is stored in 1-bit format. Binary images are repeatedly produced by a method called thresholding. This thresholding technique is applied to color images or greyscale images, and from these two we obtain the required image. It has more advantages.

6.1.4 Digital image representation

A digital image is a function f(X, Y), where X, Y coordinates are discontinuous in spatial and brightness. It is represented by a set of numbers that can be stored and controlled by a digital computer. It is formed by combing small bits of data called pixels. These pixels are stored in a computer, so the processing of the digital image is faster and less costly. The five fundamental steps in digital image processing are image acquisition, image storage, image pre-processing, communication and image display.

6.1.5 Image processing operation

Image processing is a technique applied to images to get good-quality images and is used to extract the required information from the image. Image processing operation is divided into four categories.

1. Image enhancement
    The main objective of image enhancement is to process the given input image and get a better output image which is more needful to a specific application. With this process, image edges and boundaries are improved[1]. The enhancement method does not increase the content of the data but expands the dynamic range of selected features (Figure 6.1).
2. Image segmentation
    In this method, the image is divided into several parts (segments)[2]. The aim of segmentation is to modify the representation of the image that is more purposeful and easier to analyze. It is further needful for the object recognition (Figure 6.2).

Figure 6.1   Image enhancement.

Figure 6.2 Image segmentation.

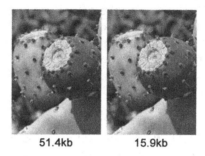

51.4kb   15.9kb

Figure 6.3 Image compression.

3. Image compression

In this method, the size of the image file is reduced without reducing the quality of the image. This reduction in image file supports the storage of more images as well as makes them easier to transfer through the internet (Figure 6.3).

4. Image restoration

This method includes cleaning up of the noise or corrupt points from the original image. In an image, corruption occurs due to many factors such as camera misfocus, motion blur and noise.

6.2 Need for color image processing

Color images include three-valued information of each pixel and measure the intensity and chrominance of the light. Before applying segmentation to the color image, it should be made precise. The proposed method identifies plant disease by processing the plant leaf image [3]. The leaf of the plant constitutes veins and disease spots with a different color. A leaf is converted to a greyscale image and then segmentation is applied to locate vein and disease spots. However, we are concerned about the disease spots not the veins [4]. To reduce vein presence, RGB component is color transformed before segmentation. Using the following three techniques color image processing is performed:

1. YCbCr color model is the most popular color space. In this model, Y represents luminance component, and Cb and Cr represent chrominance. It is widely used in video and image compression.
2. HSI color model is a tool used for developing an algorithm in image processing based on color. In HSI model, H denotes hue, which describes pure color; S denotes saturation, which describes how much pure color is mixed with white; and I stands for intensity measuring brightness.
3. CIELAB color model is defined as color space. It can be represented by mathematical operations to express the range of colors. Typically, there are three to four values of color components. It is used for display purpose and also for printing.

6.3 Feature extraction

In this method, the details of the appropriate shape information are expressed in a pattern so that the task of arranging the pattern is made easy. It is used in pattern recognition and a specific form of size reduction [5]. The process of feature extraction is helpful when you need to lessen the number of stratagems needed for processing without losing relevant information. Feature extraction can also depict the amount of surplus data for a given analysis.

6.3.1 Classifiers

Classifiers are used to identify the presence of disease-affected region in a leaf. These classifiers use few methods for detection of disease. There are three methods used as classifiers:

1. Percentage infection
2. Support vector machine (SVM)
3. Back-propagation neural network

6.3.2 Existing system

Conventionally, farmers can determine changes in the leaf color to point out the diseases by the naked eye inspection method. Generally, farmers use the naked eye to detect the diseases in plants by observation method. The detection of plant diseases is done by the trained farmer, who can detect even the small changes in the color of the leaf. The observation method to detect the disease in the paddy is time-consuming, laborious and not possible in large area agriculture fields. Best methods and practices need to be implemented for detecting plant leaf diseases, which are accurate, fast and helpful to the farmers.

Widespread diseases in paddy leaf are narrow brown-spot disease, brown-spot disease and blast disease (BD). To determine these diseases, we have various techniques. In the past decade, research using image processing was conducted to detect and analyze plant diseases [6]. Image processing techniques are aimed to decrease the subjectiveness and enhance the throughput to detect paddy diseases. The detection of the diseased part of the leaf is basically done in two steps. The first step is image acquisition and the second step is segmentation using spot detection method and boundary detection. The classification of diseased paddy leaf is based on zoom algorithm. The zoom algorithm works on self-organizing map neural network.

We have two processes to build self-organizing map of the input vector. The first process uses the zeros to padding, and the second process uses the detection of missing points in interpolation. The interpolation method is used to normalize the spots by zoom algorithm. To apply the zoom algorithm for the segmentation, the first step is image acquisition and the second one is the K-means clustering method. Color co-occurrence method is applied to analyze the infected leaf and stem. Finally, the algorithm of back propagation is used with the neural network for classification diseased plants [8]. Using image processing methods, the identification and categorization of plant diseases is done with accuracy of around 94.00%.

Some other techniques for determination of the disease constitute image acquisition, color processing, feature extraction and classification by considering the production rule technique with a forward chaining method. Precision is around 92.00% for this method, and it is not a recommended technique for disease identification.

6.3.3 Suggested method

In this method, the identification of plant disease is done by internet of things–based technique. In this technique, sensors are placed at different places across the field, these sensors collect data and send it to a processing unit where images are processed with different methods to identify the disease-affected plants by examining the color of the leaf. This process is done by feature extraction and machine learning techniques. The suggested method is color image processing, along with three different techniques: (i) image segmentation, (ii) image smoothing, (iii) feature extraction. We got 96% accuracy in disease classification by using the vector machine demonstration. Our approach is towards massive classification of plant diseases with automated techniques.

6.4 Motivation and objectives

6.4.1 Motivation

From literature survey we get an idea that existing technique is less precise in identifying diseases in a plant, so it is necessary to find and implement a new method. Hence, segmented image using finest classifiers are applied to increase the accuracy.

6.4.2 Objectives

1. Pertaining HIS color processing method
2. Eliminating noise by modifying median filter
3. K-means clustering used to measure the performance of segmentation
4. SVM classifier for categorizing disease and image feature extraction

6.5 Proposed approach

To overcome the drawbacks of the previous approach, a new technique must be implemented to increase precision in disease detection with less computations. For that,

K-means clustering and SVM techniques are combined, so this is a hybrid technique. With this technique, accuracy can be achieved to a desired value.

6.5.1 Image acquisition

Image acquisition is obtained from a camera. It is the primary process in digital image processing. Basically, it involves various techniques like scaling, eliminating noise and enhancing the contrast of the image. The subjected leaf images are captured from highest resolution cameras with more pixel values such that obtained image can be processed with various techniques to extract the desired area of interest from the leaf and easily identify the disease-affected leaf.

6.5.2 Image enhancement

In this method, digital images are modified such that they become more acceptable images from the previous ones. With the help of image enhancement more study can be done on the obtained images, for example, detecting noise in the image, improving the brightness of the image and extracting required area from the image.

6.5.3 Color image processing

Color image processing is very important for use of large digital images over websites. This includes image processing and color modeling by digital processing. In this method, HSI color processing is applied to the image to get enhanced image (see Figure 6.4).

Based on the color perception, the color model device is chosen as HSI. In the HSI color device model, H represents hue and illustrates the real color, S represents the saturation and indicates the luminosity and I represents the intensity. It points out the amplitudes shown in Figure 6.4.

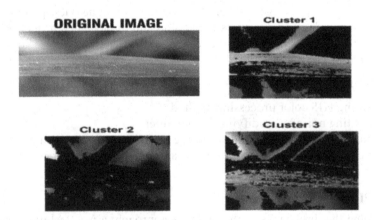

Figure 6.4 RGB to HSI conversion.

6.5.4 Image segmentation

In this process, the image is divided into smaller parts called segments to make the analysis and further processing easy. An image is subdivided using segmentation. The segmentation of the image divides it into objects or regions. The subdivision of the image depends on the depth of the problem. Segmentation algorithms work basically on two properties of the image: similarity and discontinuity. The first category defines sudden changes in the intensity of the image. The second category is based on dividing the image into objects. Segmentation algorithm mainly implemented on plant leaves K-means clustering.

6.5.4.1 K-Means clustering

K-means cluster process is applied to the image to do cluster analysis. It is the best method to solve problems related to clustering of the best learning algorithms that solve the clustering problems. K-means algorithm is used to detect the iterative K-centers. The sum of distances of the data points is optimized from their centers. In many applications, the K-means algorithm of clustering is used to segment the paddy leaf image into many clusters of disease. Figure 6.5 shows the output of the K-means cluster algorithm for a leaf infected with the disease.

6.5.5 Feature extraction

The size reduction of the image data is done by feature extraction. The segmented region is very easy for doing study as it is filled with suitable colors and shapes. Most of the features are used to illustrate an image of paddy leaf.

6.5.5.1 Percentage of leaf area infected (IA)

Using the methods as indicated earlier, the image is characterized and segmentation detects the diseased area of the leaf. By using below formula, we can calculate

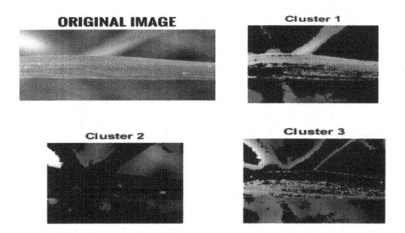

Figure 6.5 Segmentation of healthy leaf using k-means clustering.

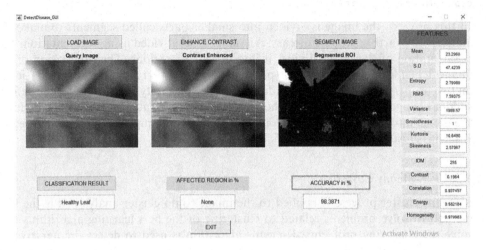

Figure 6.6 Healthy leaf features extraction.

percentage of the infected area of the leaf. In the formula, DA is the diseased portion of the leaf area and LA is the total leaf area (Figure 6.6):

$$IA = \frac{DA}{LA} \times 100.$$

6.5.6 Classification of image

We did image classification using classifiers. Based on the degree of disease in a leaf, a classifier is used for classification and classification depends on the features of the last step. In machine learning it is used to divide values into two groups: one of positive values and other of negative values, depending upon certain properties. Support vector is used for identification and classification of an image.

6.5.6.1 Support vector machine (SVM)

In advanced machine learning, the classification is done by SVM. The SVMs aim to create a decision boundary using which we can identify boundary from two suitable output values. SVM is also used for classification purpose. It is a supervised algorithm with which we can the change the boundary position also. The points closer to the boundary are called support vectors, these are required for further analysis.

6.5.6.2 Image acquisition

Using different digital devices like digital cameras, mobile phones and laptop, we can capture the image of the leaf. Generally, image can be greyscale or color. Six images are considered standard as per the dataset. Out of six, three are colored and three are greyscale images. Our first task is to check whether the image is a greyscale or color image. If the image is in RBG model of the color, first the image will be converted into HSV image and then greyscale image.

6.6 Implementation of the proposed method

Flow chart of the proposed method.

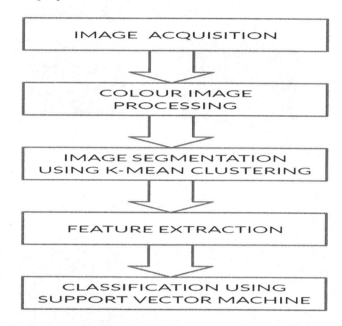

6.7 Segmentation simulation result

Simulated output of classification, image processing and segmentation of the infected paddy plant leaf is shown in Figures 6.7–6.10.

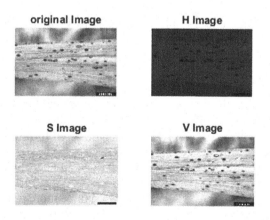

Figure 6.7 By using median filter noise removal.

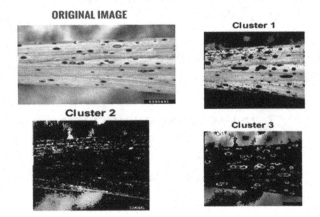

Figure 6.8 Using k-means clustering image segmentation is done.

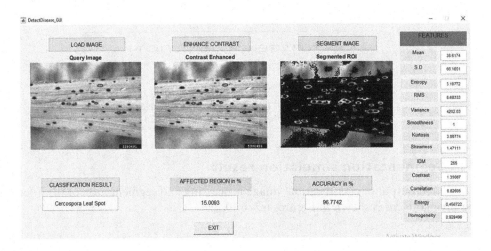

Figure 6.9 Extraction of desired features from the image.

6.8 Advantages

1. At one place many different crops can be analyzed and there is no need of different methods for different crops.
2. This system is easily accessible by the farmers; by adding some more features with Graphical user interface (GUI).
3. This system is scalable, so we can scale up this system as per the requirement.
4. This system is adaptable.
5. There is a huge future scope, so this system will be more powerful in future.
6. It is a fast and accurate approach to identify plant diseases.
7. It is not a laborious process.

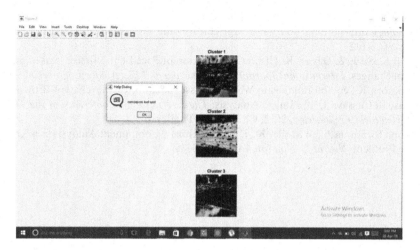

Figure 6.10 SVM method for identification of the disease.

6.9 Conclusion

This chapter presents the detection, diagnosis and identification of diseases in the paddy leaf. The suggested method can be followed at each step for the type of crop and its leaf. The most important task is extraction of the features of the leaf and doing further analysis. The classifier is used to set the tone for proper results. The classifier-based neural network is one of the best methods. The method described in this chapter can be applied in real-time applications. This system uses color-based segmentation. A lot of research is going on to detect infected crops automatically to get good yield.

6.10 Future scope

Using above algorithms and the literature on detection of infected leaf is not sufficient. The future scope is that more efficient algorithms should be defined to find plant diseases accurately as the above algorithms are not equally applicable to all crops.

The noise removal from the image using the above algorithms is most liable and recognizes the plant disease. The above algorithms are very efficient and fast in computation to detect the plant disease.

Bibliography

Asman & Verma R. K. (2018) *Identification of Paddy Plant Disease Using Image Processing.* India Accenture Services Pvt. Ltd., Pune, Maharastra, India.

Chaudhary P., Chaudhari A. K., Cheeran A. N. & Godara S. (2012) Color Transform Based Approach for Disease Spot Detection on Plant Leaf.

Kaur J., Singla S., & Singh R. (2016) Classification of Segmented Image Using Increased Global Contrast for Paddy Plant Disease International Journal of Latest Trends in Engineering and Technology Vol.(7)Issue(3), pp. 419-426 DOI: http://dx.doi.org/10.21172/1.73.555

Kim W.Y. (1994) A practical pattern recognition system for translation, scale and rotation Invariance. *International Conference on Computer Vision and Pattern Recognition*, pp. 391–396.

Nisale S. S., Bharambe C. J., & More V. N. (2011) Detection and Analysis of Deficiencies in Groundnut Plant Using Geometric Moments. *World Academy of Science, Engineering and Technology* International Journal of Agricultural and Biosystems Engineering Vol:5, No:10, 2011.pp 608-612

Phadikar S., Sil J., & Das A. K. (2012) Classification of Rice Leaf Diseases Based on Morphological changes. *International Journal of Information and Electronics Engineering* 2 (3).

Powbunthorn K., & Abudullakasim W. (2012) Assessment of the Severity of Brown Leaf Spot Disease in Cassava Using Image Analysis. *The International Conference of the Thai Society of Agricultural Engineering*, 2012, Chiangmai, Thailand.

Scholkoph B., Smola A., & Muller K., (1998) Nonlinear Component Analysis as a Kernel Eigen Value Problem. *Neural Computation* 10, 1299–1319.

Chapter 7

IoT in agriculture

Survey on technology, challenges and future scope

Seema Nayak

IIMT COLLEGE OF ENGINEERING

Manoj Nayak

MANAV RACHNA INTERNATIONAL INSTITUTE OF RESEARCH AND STUDIES

Govind Singh Patel

LOVELY PROFESSIONAL UNIVERSITY

7.1 Introduction

The internet of things (IoT) has plentiful applications in different fields. The initiative of IoT has shown a novel track of advanced research in agricultural dominion which is capable of offering many solutions toward the modernization of agriculture. Data are captured by mobile devices and sensors; for broadcasting data radio frequency (RF) technologies are used. To be persistent and functional in an effective way, sensor networks use verification and optimization methods. The processes and predictive analytics can be automated using big data approaches and increase many activities in the real time. This chapter covers the IoT technologies used for developing systems that support different agricultural processes, its applications for agriculture, challenges that IoT faces and future scope for upgradations of IoT in agriculture segment.

7.2 Innovative research in agriculture

Due to the new state of shrinking water levels, rivers and tanks are drying up and there is an urgent requirement of proper utilization of water. Crops can be supervised by placing proper sensors at right locations. The critical investigation survey in the area of agricultural shows that the income from agriculture is decreasing day by day. The implementation of technology in the area of agriculture plays a dynamic role in increasing the production and reducing the man power [1]. Current scenario in agriculture as well as in civilization demands the production of food in better way for the universal population. To improve the net production various new technologies are being used in agricultural field which provide an optimal solution for collecting and processing information [2]. World economy is increasing due to the qualified human capital and improvements in science and technology. Due to this increasing development in smart farming methods, farmers can remotely monitor the crops using sensors. IoT is a network of interconnected devices which transfer data effectively without participation of humans. Nowadays farming sector uses IoT devices and smart phones [3].

The new sensors-based computer applications allow to obtain the correct information regarding the soil, crop and climate in contrast with the old methods. Such methods are used to improve the quality products, practices and also the raw materials used in the process. Hence smart agriculture application of IoT is more competent as compared with old-style methods as shown in Figure 7.1. It helps to reduce fertilizers, wastage of water and improves the crop harvest [1]. It involves modern information and communication technologies application for clean and ecological growth of food for the masses [4]. Thus, IoT is remodeling the agroindustry by providing the agriculturists with broad range of strategies [5].

India is a developing country and the GDP growth contributed by the service sector is 57.9%, industrial sector contributes 24.2% and agriculture sector contributes 17.9% [6]. Precision farming deals with data collection, application and data interpretation. It is the best farming approach among various farming methodologies. The entire process of precision agriculture is governed by the GPS technology, electronics and

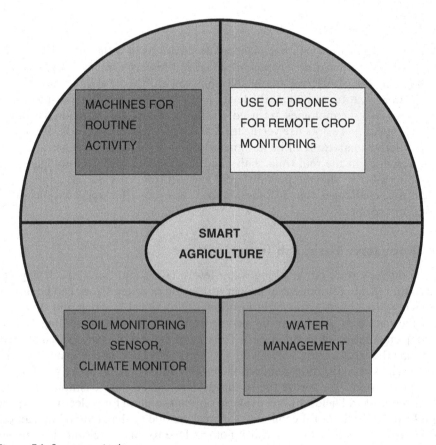

Figure 7.1 Smart agriculture.

network transmission. But if there occurs any climatic or natural disaster, this farming system will become a complex system.

7.3 IoT for agriculture

This section elaborates the role of IoT in agriculture and the efforts of related research and its application of IoT technologies. The critical survey is used to guide the use of such new ideas in urban agriculture and in precision agriculture.

7.3.1 Internet of Things

The term "Internet of Things" comprises the technology developed in 1999. After that IoT produced a lot of interest in various fields. It is expected that the number of things which are connected to IoT will grow from 20 billion in 2015 to an estimated 200 billion by 2020. In real life, things are connected to internet and data can transfer over the cloud. The design of IoT is shown in Figure 7.2. A number of functional blocks of

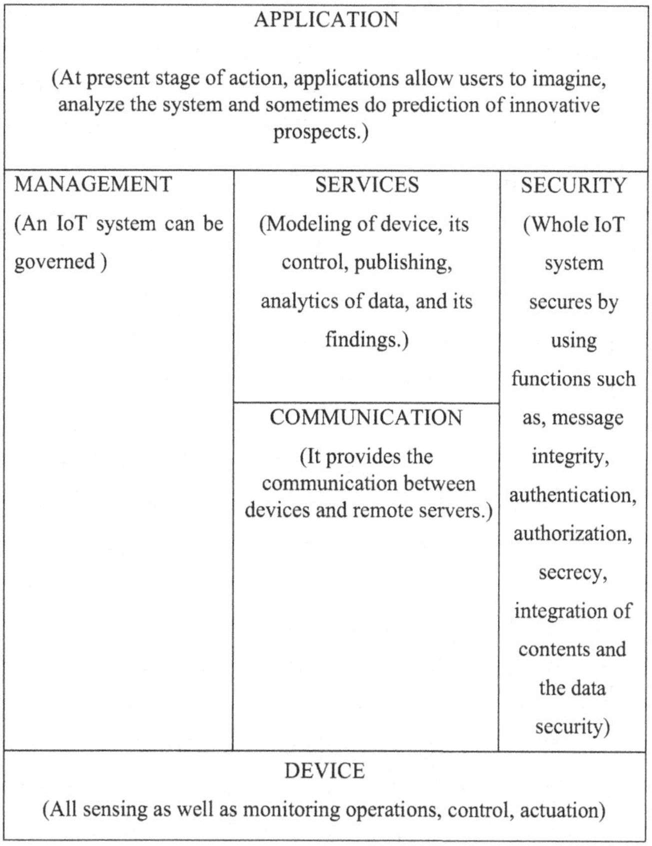

APPLICATION (At present stage of action, applications allow users to imagine, analyze the system and sometimes do prediction of innovative prospects.)		
MANAGEMENT (An IoT system can be governed)	SERVICES (Modeling of device, its control, publishing, analytics of data, and its findings.)	SECURITY (Whole IoT system secures by using functions such as, message integrity, authentication, authorization, secrecy, integration of contents and the data security)
	COMMUNICATION (It provides the communication between devices and remote servers.)	
DEVICE (All sensing as well as monitoring operations, control, actuation)		

Figure 7.2 Design of IoT system.

IoT system are used to simplify several functions of the system like sensing, services, management and communication [2].

7.3.2 IoT devices

The network capabilities are the core features of the IoT devices, therefore many machine-to-machine (M2M) communication takes place [7]. IoT-enabled devices can interchange data over connected devices and applications, and further can gather data from various devices and processes either locally or send it to centralized servers or cloud. IoT systems can deliver services like modeling, device controlling and analysis of data. Management block in an IoT device manages the system. In the functional block diagram, security block protects the system by authentication and authorization. At present stage, applications block allows users to imagine and analyze the system status and predict the future. The structure of IoT devices is shown in Figure 7.3.

IoT applications in the field of agriculture shield a large number of circumstances. Researchers [8] set applications in networks of scalar sensors, which are used for sensing and control of agricultural infrastructures like greenhouses, the distant image capturing and the detection of insects using multimedia sensor networks and detection of plant diseases by processing, product tracing and remote identification using tag-based networks (e.g. Radio Frequency Identification, Near Field Communication). Crops or other problems in farmlands, whose positions may change, cause significant interference in the communication between nodes [9]. So, low-power, embedded devices are used for communication due to their long-standing stability. The number of peripheral devices like sensors and actuators to be used is determined by the number of digital as well as analogue inputs/outputs.

IoT devices have self-configuring capability, which allows a large number of devices to work together to provide certain functionality like weather monitoring.

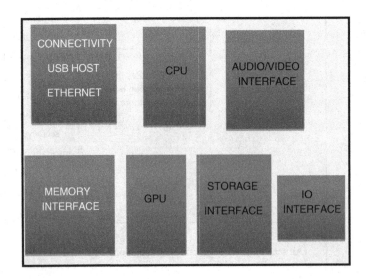

Figure 7.3 Structure of IoT devices.

These devices have the ability to design themselves, set up network and upgrade themselves to latest software with minimum user intervention. IoT devices support a number of communication protocols to communicate with other devices. Each IoT device has a unique identity and unique identifier such as IP address or Uniform Resource Identifier. In association with the control, configuration and management infrastructure, IoT devices permit users to appeal the devices, supervise their status and manage them remotely. Intellectual decision-making capability of IoT devices increases the overall network energy efficiency, and therefore the network lifetime increases.

7.3.3 Benefits of IoT in agriculture

The role of IoT in agriculture as discussed in several literature reviews is as follows: Agriculturists take advantage of software and hardware resources and large amounts of data in rural, semi-rural and urban zones. Decision based on real-time data are useful in minimizing costs and several inputs and can gauge the qualitative production of food. Direct relationship with the consumer can be established by application of a suitable business model. Technology-driven crop monitoring simply brings down the costs as well as the theft of machinery used in farming. Sensors are used in automatic irrigation systems to monitor humidity, temperature and soil moisture values for further processing and analysis. Decision support systems scrutinize a huge amount of data to overall improve the productivity and enhance the operational efficiency. Hence, technology-driven agriculture increases the productivity by decreasing the labor-intensive manpower [10], and research suggests that it also increases the yield as well as farmers' welfare and financial conditions. Complete automation in irrigation system helps in optimization of water consumption in agriculture [1], as wireless sensor networks are used to acquire data about several factors like humidity, moisture and temperature from several locations of the field.

7.4 IoT-based agriculture application

The technology plays a vital role in agriculture. IoT-based agriculture is shown in Figure 7.4, having different layers from physical layer to user-experience layer. This system performs acquisition of data, statistical analysis, storage and then decision-making on available data.

Developments in sensor technology and other electronic devices reduced the costs and have added manifold advantages to traditional agriculture practices. Such evolution of conventional method of agriculture to precision and micro-precision agriculture [9] lately brought a phenomenal improvement in agriculture productivity. IoT application in agriculture is about providing farmers with the automation technologies and decision tools which seamlessly incorporate products, knowledge and services for better productivity, quality and profit [11].

IoT has found its application in several areas, such as

- Proper utilization of pesticides and fertilizers which helps in rising the crop quality as well as reducing the cost of the farming.
- Water usage in farming and related activities; improved irrigation management system is used nowadays.

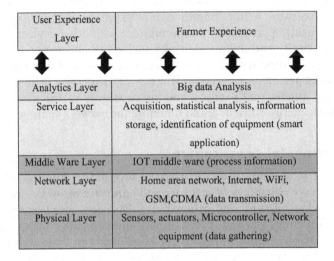

Figure 7.4 IoT-based agriculture.

- Wireless sensor technology and communications helped in water quality monitoring on parameters like temperature, pH, turbidity, conductivity, dissolved oxygen.
- Monitoring of herd of cows and other animals grazing can be done using IoT.
- Agriculture and greenhouse are related to each other. Increase in climate temperature due to greenhouse gases and its effect on agriculture is monitored.
- UAV and drones are used for fruitful farming.
- Knowledge of soil is important for the production of corps, monitoring of soil is done using IoT.
- Supply chain management is needed for gaining profits which can be monitored by IoT.

7.5 Challenges

There are several challenges with use of IoT in agriculture:

- Seizing tremendous volume of generated heterogeneous data by sensors and performing the analysis for a large number of data for study of crops.
- Cost of hardware and software to be set minimum while maximizing the system output.
- There is a challenge of integrating historical data of crop with IoT devices.
- Agriculture-related standardization must be done in IoT and should strictly be followed.
- IoT requires the use of a common platform to be successful in agriculture.
- Agriculture systems driven by IoT based on non-conventional source of energy harvesting solutions must be considered.
- Suitable programming tools and low-power capabilities are needed in agriculture so as to keep power consumption within the budget of small- and medium-power harvesting modules.

- Robust and reliable technologies are used to transfer data to meet the challenges of the rural environment as desired.
- Authenticity and confidentiality to secure data integrity during the process of data acquisition with key management protocols needs to be ensured by adopting routing policies and authentication policies.
- Collaborative effort from all stakeholders is required to introduce universal standards in developing IoT, keeping aside the competitiveness.

7.6 Conclusion

Agriculture remains the backbone of an economy, and IoT-driven agriculture is proposed for the progress of agricultural and farming industries, which will contribute to the national GDP. IoT technologies acquire information on climate, humidity, temperature and soil fertility for remote monitoring of crops using wireless communication and sensor technologies. Technology-driven agriculture has enriched the farmers with more financial and intellectual capability to face the agriculture paradox. Even farmers who are illiterate or less educated can make use of the IoT to leverage the crop production. Literature reviews suggest multifold benefits of using IoT applications in agriculture and farming with some challenges. Hence there is a dire need to carry forward the research on application of IoT in agriculture and face the challenges. There is need for more investigation to achieve ecological cultivation of food. To sum up, the various aspects of IoT technologies for smart agriculture are discussed in this chapter.

7.6.1 Future Scope

In the future, IoT-based agriculture systems can be improved:

- using deep learning technology to make the system fully automated [12]
- using low-cost, portable devices and systems and IoT-based agriculture
- using user-friendly design and ergonomics appropriate systems
- using green computing techniques to disseminate with the IoT-based agriculture
- using suitable devices for communication with the external world
- using implementation of artificial intelligence and machine learning techniques
- by less human intervention and low maintenance system
- by keeping in mind the maintenance time and cost, IoT systems should be designed
- by having robust and fault-tolerant architecture agriculture system
- by avoiding unwanted situations through equipping farmers with local weather sensors which can communicate with national weather centers

References

1. T. Vineela, J. NagaHarini, Ch. Kiranmai, G. Harshitha and B. AdiLakshmi, "IoT based Agriculture Monitoring and Smart Irrigation System Using Raspberry Pi", *International Research Journal of Engineering and Technology*, 2018, 5 (1), 1417–1420.
2. P. Pratim Ray, "Internet of Things for Smart Agriculture: Technologies, Practices and Future Direction", *Journal of Ambient Intelligence and Smart Environments*, 2017, 9, 395–420.

3. N. Udhaya, R. Manjuparkavi and R. Ramya, "Role of IOT based Indian Agriculture Sector", *International Journal of Advanced Research in Computer and Communication Engineering*, 2018, 7 (3), pp. 84–86.

4. B. S. Shruthi, K. B. Manasa and R. Lakshmi, "Survey on Challenges and Future Scope of IOT in Healthcare and Agriculture", *International Journal of Computer Science and Mobile Computing*, 2019, 8 (1), 21–26.

5. M. Naresh and P. Munaswamy, "Smart Agriculture System Using IoT Technology", *International Journal of Recent Technology and Engineering (IJRTE)*, 2019, 7 (5), 98–102.

6. K. Kishore Kumar, B. Godi, D. Manivannan and A. S. Muttipati, "Role of IoT in Enhancing Agricultural Techniques", *International Journal of Innovative Technology and Exploring Engineering (IJITEE)*, 2019, 8 (7), 332–334.

7. R. Shete and S. Agrawal, "IoT based urban climate monitoring using raspberry Pi." *International Conference on Communication and Signal Process*. April 6–8, 2016, India IoT. 2008–2012 (2016).

8. J. M. Barcelo-Ordinas, J. P. Chanet, K. M. Hou and J. G. Vidal, "A survey of wireless sensor technologies applied to precision agriculture", July 2013, *Conference, 9th ECPA Conference*.

9. A. Tzounis, N. Katsoulas, T. Bartzanas and C. Kittas, "Internet of Things in Agriculture, Recent Advances and Future Challenges", *Biosystems Engineering*, 2017, 164, 31–48.

10. R. Gómez-Chabla, K. Real-Avilés, C. Morán, P. Grijalva and T. Recalde, "*IoT Applications in Agriculture: A Systematic Literature Review*", Springer Nature Switzerland AG, 2019, pp. 68–76.

11. O. Elijah, T. A. Rahman, I. Orikumhi, C. Y. Leow and M. N. Hindia, "An Overview of Internet of Things (IoT) and Data Analytics in Agriculture: Benefits and Challenges". *IEEE Internet of Things Journal*, 2018, 5, 3758–3773.

12. B. Sidhanth Kamath, K. K. Kharvi, A. Bhandary and J. Elroy Martis, "IoT based Smart Agriculture", *International Journal of Science, Engineering and Technology Research*, 2019, 8 (4)pp. 112–116.

Chapter 8

Role of IoT in sustainable farming

Rajasekhar Manda

ADITYA ENGINEERING COLLEGE (A)

P. Rajesh Kumar

AU COLLEGE OF ENGINEERING (A)

8.1 Introduction to Internet of Things (IoT)

Computing and communication have been revolutionized with the invention of Internet of Things (IoT). This term was coined by Kevin Ashton from Proctor & Gamble Consumer Company in the year 1999. Initially it was confined to manufacturing and business only, and also known as machine to machine (M2M). It uses the internet and connects with unique features through several devices. By 2025, 41.6 billion devices might get connected with IoT. These incorporated devices could be electricity meters, Bluetooth headsets, sensors, water system pumps, control circuits, etc. However, the list is not just confined to these devices only, but it covers vast applications. The raw data are collected from devices such as sensors and gadgets. The information gathered is handled, ordered and contextualized to the application level through brilliant execution mechanism.

In the early days, radio-frequency identification (RFID) was an essential part of IoT.

8.1.1 Fundamental characteristics of IoT

The information collected by IoT devices undergoes a five-phase life cycle.

Create phase: Different sensors and devices collect the information from the physical world.

Communicate phase: The collected data are sent through the network to the desired destination.

Conglobation or aggregate phase: The data are summed up by the devices.

Evaluation or analyze phase: The aggregated data are used to generate the basic pattern for optimization and control actions.

Activity phase: The designated action is performed based on the data available.

IoT involves different characteristics such as connectivity, dynamic nature and large scalability, heterogeneity, intelligence and security (Table 8.1).

8.1.2 Internet of things in agriculture

The ultimate aim of technology-based agriculture is to cultivate in a smart manner, produce the farm with less wastage and deliver the produce to end-user at reasonable

Table 8.1 IoT characteristics

IoT characteristics	Explanation
Connectivity	Consistent objects are united to enable the network in IoT.
Dynamic nature	The collected information undergoes dynamic changes progressively with time, locality and human intervention, etc.
Large scale	The number of devices connected together and managed is much larger compared with devices connected to the internet. Stated 5.5 million devices connected to IoT. About 30% of device utility increased in 2016.
Heterogeneity	It is the main characteristic of IoT. The devices interface through different networks based on hardware flatform. Heterogeneity of devices and environment implies their interoperability, adjustability, scalability and separateness.
Intelligence	IoT works well with any hardware, software, programming and algorithms. This facilitates the users to interact with a smart device in intelligent manner with GUI.
Sensing	IoT cannot work without sensors in many applications as interaction with environmental data is very necessary. The sensing data are input from the physical world using sensors.
Security	IoT devices are very sensitive to security threats. The data transferred through all the network layers have to be protected.

Source: https://www.educba.com/iot-features/

price and quality. Nearly 50% of the farm produce could not reach the consumer due to many reasons like wastage, transportation cost, climatic conditions, lack of direct marketing etc.

IoT system in agriculture works on the real-time data that can be collected by the farmers (humidity, temperature, soil moisture, soil pH value, nutrient applications, etc.) through different sensors and is used to take action when abnormal situations arise. This reduces the stress on farmers in continuously monitoring the fields. Along with this, there are several benefits of using IoT in agriculture. Right from the soil testing and farm reaping enormous devices exist to adopt new techniques in farming.

1. The collected data can be analyzed to predict the future statistics.
2. Risk management can be achieved through smart sensors and networks in farming.
3. Complete automation without human intervention helps to improve productivity and activities are performed in an efficient manner.
4. Good quality can be achieved through different farming techniques and automation methods.
5. While harvesting, grading automation and packaging can be done through proper identification and tracking techniques.
6. Based on past data farmers can get the advice on optimal prices for their crops throughout the year (Figure 8.1).

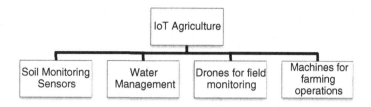

Figure 8.1 IoT in agriculture.

8.1.3 IoT-based crop monitoring

Connecting different types of devices together to form a huge network is the main goal of IoT. Three aspects can be accomplished through IoT: automation, communication and cost-effectiveness. IoT emancipates people from different farming activities; thus, it helps in saving their time and make them more productive. Environment monitoring provides data about air, temperature, condition of the soil, prediction of rain and water quality, and crop monitoring and animal monitoring enhance the productivity of the farm.

Wireless sensor network (WSN) with ATMEGA2535 processor and IC-S8817 BS with ZigBee protocol observes the conditions of the farm along with a web application to store the data. Greenhouse-based intelligent monitoring system based on ZigBee protocol performs transmission and receiving, data processing and data acquisition functions. The main aim is to reduce the manpower and cost, provide remote intelligence, and switch from wire to wireless technology, using B-S structure and cc250 as processing chip for coordination (Dan *et al.*, 2015; Haule & Michael, 2014).

Factors like soil moisture, soil pH, leaf wetness, and atmospheric pressure trigger sensors working on sprinklers. This automatic sprinkling system is implemented using WSN for conservation of water. The system also informs the farmer of the soil pH level through SMS using a GPS model (Vijayakumar & Nelson Rosario, 2011; Zhang *et al.*, 2010). An effective solution is provided for household shrimp farming with improved accuracy and monitoring of environmental conditions using graphical user interface (GUI) programmed with LabView. This reduces the cost of labor and power consumption (Duy *et al.*, 2015; Sivasankari & Gandhimathi, 2014).

A smart sensor enables the embedded system to make interactions with the environment that forms WSN. These sensors gather the soil moisture and temperature data applied to solve monetary and environmental problems. Data and information are provided using cloud computing to improve data storage capacity and processing capability (Sales *et al.*, 2015).

This IoT-based crop monitoring enables the collection of data about crops through wireless networks and the farmer uses this information to make the right decisions without bringing any loss to the crops. But wireless sensors pose challenges like having a user interface and less power consumption to improve productivity.

8.1.4 Applications of IoT

The applications of IoT can be found in technology-efficient artificially intelligent (smart) devices such as electronic gadgets (smart watches) in almost every area

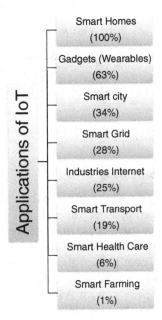

Figure 8.2 IoT applications.

including smart cities and homes, transportation, industries, power sector, health and agriculture. The usage of IoT in terms of percentage in different sectors according to Google searches (valid till 2014) is shown in Figure 8.2.

8.2 Introduction to smart farming

Recent research indicates that IoT has changed the majority of agriculture industries for a better future. With the extreme climatic changes and increase in population, there is a demand for more food to meet the requirements. The IoT-based smart farming enables the farmers to apply the smart technologies to the fields so as to reduce waste and improve the productivity. This smart farm revolution refers to the usage of latest technology inventions in agriculture to enhance the quality and quantity in crop harvesting as shown in Figure 8.3.

IoT-enabled sensors provide the real-time data to the farmers to take the right decision for better yield. For example, use of drones in spraying the pesticides reduces the labor cost and soil monitoring enhances the productivity. The government is encouraging the farmers to use the latest IT-based farming equipment by giving subsidies (Mooney, 2017). To preserve the technology revolution the skilled and semi-skilled manpower is trained through technical vocational education and training. This creates the employment opportunities for the people.

Smart farming can be achieved by introducing automation in irrigation, monitoring and controlling of crop growth, pest control and insect population control and adopting the latest IoT-based technologies.

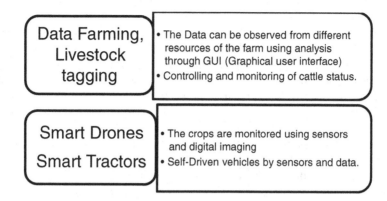

Figure 8.3 IoT smart farming.

Smart farming includes environmental data collection, monitoring and controlling. Traceability systems improve the customer's faith in the agricultural products. Smart farming enables the embedded sensors to collect the environmental data. Hydroponic growing system is a system meant for feeding the plant without soil using mineral nutrient solutions. In this system, humidity, temperature of the water, electrical conductivity and potenz ("strength" in German) of hydrogen pH play the main role in growing plants.

It includes two subsystems: foggy spray system (for regulating the temperature and humidity) and auto-dosing system (enables to measure the water conditions using pH, electrical conductivity).

8.2.1 IoT smart farming policies in different countries

Many countries are trying to adopt advanced monitoring techniques facilitated through IoT. However, IoT opportunities are stopped by the governments of different countries as they are investing more on crop productivity improving methods.

India: Agriculture is the backbone of India. IoT policies are focused on agriculture to strengthen the Indian economy all over the world. In 2015, the ministry of communication and information technology released a policy on IoT which states that IoT has the potential to automate the agriculture industry and transform the digital landscape (http://meity.gov.in).

Australia: In Sydney, the government of Australia invested over AU$134 million to strengthen the farming and IoT technology for smart farming. Due to privacy and security reasons in 2015, American Bureau took over the farm data (http://www.agriculture.gov.au/).

China: To enhance the profitability using IoT in the field of agriculture the 13th five-year plan was launched by China in 2015. Multiple products and technologies, including 426 applications in 8 provinces, were launched by a project. To transform the agriculture in a more efficient and innovative way, Huawei Company launched an App named NB-IoT. It provides cost-effective agriculture solutions without using gateway implementations like cellular network. Due to its special features like wide area coverage, a number of connections were able to solve the agriculture data issues (http://mit-insights.my/).

Malaysia: Two policies were made in this country before and after independence, that is, one during 1948–1957 and another during 1957–2020 (http://ap.fftc.agnet.org). The main theme of the policies was to eradicate the poverty and enhance crop productivity. Malaysian Institute of Microelectronic System (MIMOS) developed several solutions for developing the agriculture. To collect the environmental data a sensor named Mi-MSCANT pH was developed. The MIMOS developed a framework to integrate the traders, suppliers and unified agriculture producers. WSN and microelectro mechanical system are able to collect the environmental data automatically (http://mit-insights.my/).

USA: Millions of dollars have been invested in generating the new agricultural technologies to meet the requirements of food and energy. National Institute of Food and Agriculture created a sensing technology for smart farming practices through the project named "Internet-of-Ag-Things". This project is based on precision technology to improve the agriculture industry's efficiency through utilizing fertilizers, organic food and water resources (https://reeis.usda.gov). A dataset was developed by the U.S. Department of Agriculture for water management challenges and issues that are affecting the agriculture. The experts are using this dataset to develop the solutions for different agricultural problems.

Thailand: The National Electronics and Computer Technology Center of Thailand focuses on agriculture products like cassava, rice, rubber and sugar by applying latest IT technologies (https://www.nectec.or.th/en/). The main aim is to help the rural farmers of Thailand to increase the agriculture products (DTAC Debuts, 2019). An IoT-based irrigation control system was developed by Faculty of Science and Technology of Thailand University to present advanced water cycle timing. In 2015, a smart farm service named Farm D Asia was launched to boost the productivity of agriculture along with leadership products. The same department developed a pesticide drone that covers up to eight acres of land in its one flight. Agricultural System Integrator program was launched by the National Science and Technology Development Agency for farmers to make the smart farms successful.

France: The Agriculture Innovation project 2025 includes the ministry of agriculture (France). This project's goal is to create the incubators for improving agriculture fields, monitoring the climatic conditions and strengthen the agricultural land. The ministry gave its permission to share the data with farmers to find new solutions for agriculture problems in the fields (http://agriculture.gouv.fr). The Common Agriculture Policy (CAP) gives support to French agriculture monitored by European Union. Different agro projects on ecology were included in the next 5-year plan in the new CAP frame.

Ireland: A program was developed by Irish Farmer's Association especially for the farmers to make them aware of water and energy conserving methods and soil fertility improvement technologies (Ojha et al., 2015). Over 8700 euros were saved by the companies. Besides, there was an increase of 47% in profit due to soil fertility and 10% drop in greenhouse gas emission and 21% of pasture management was achieved by following the instructions given to farmers. In 2016, Ireland VT-Network launched SigFox network to provide a solution for farm asserts, tracking security sensors (https://vt-iot.com/vt).

8.2.2 Agriculture-based smartphone applications

Several domains in agriculture can be automatically monitored to enhance productivity of the agriculture with less human effort. The sensors-based equipment that are connected to IoT can be operated using software-based applications. These applications interact with the humans using GUI. These apps include the farm management, animal stock, weather report, goods availability and soil parameters. These applications include mainly three monitoring and management systems (Table 8.2).

All these apps work on Windows, Android, iPhone, iPad, and BlackBerry operating systems.

Table 8.2 Smartphone-based agriculture apps

Smartphone agriculture app types	App names
Agriculture information management	Ag Data: To record the experiments done on crop cutting digital data; developed by NIC, India. GSM: A grain storage management app. Kisan Suvidha: Farmers' welfare app. Cow: For best practices in animal husbandry through text messages. Virtual Farm Manager: Edit all the fields ahead to sowing a variety of seeds. Pasha Pusan: To protect and develop the dairy farms. Satellite field monitoring app. Pusa Krishi: Aims to inform the farmers on latest technologies. Kheti-Badi: To encourage organic farming.
Weather information	Skymet Weather: Location-based weather forecasting over 7500 locations in India. Agroclimate: Advanced solutions for climate and weather. AgruWeather: Worldwide best reliable data for farmers. WeatherBug: Warning-based weather forecast. IFFCO Kisan: For rural empowerment. It provides access to weather forecasts, current market prices, one-touch facility to consult with agricultural experts, latest agricultural advisory, access to library of best practices for different crops and much more.
Calculators	Pantheon: Pesticide calculator. Agfleet: Decision-making app for productivity improvement. Agri Logix: Time and expenses and saving calculator. Other blend: Fertilizer calculator. ROI: Return on investment calculator.
Keeping the records	FARMapper: To maintain irrigation records.
Precision farm	Agrible: Accurate and location-based rainfall measurement. FieldNET: Control and monitor fields. ArcGIS: Query the maps and distance. Field Navigator: Hassle-free field monitoring in poor-visibility conditions.
Soil sampling	SAGBI: Soil agriculture ground water banking index. Yara Image IT: Nitrogen uptake by the crops. SoilWeb: GPS-based soil type measurement.
Area maps	FARMapper: Farm maps developed effectively.
Goods pricing	KrishiMitr: Provides latest commodities and fertilizers pricing list.
Composition mixing	PotashCrop: Online agriculture solution for farmers. Tank mixing.
Financial service	Agrisync: Connects the farmers with industrialists and other supporters.

Table 8.3 Efficiency of milking types

Serial number	Milking type	Pictorial representation	Efficiency (cows/hour)
1.	Bali		50
2.	Swingover		60
3.	Herringbone		75
4.	Rotary		250

8.2.3 Smart dairy farming

The dairy farming is milk production agriculture for long term which is an alternative food for human infants. Smart dairy farming includes three main areas:

1. Automatic milking
2. Automated feeding
3. Quality and health of cattle

The automatic or robotic milking has existed in the industry since 1990s. Cow feeding with different techniques is shown in Table 8.3.

The smart electronics, cameras and laser technology are used to identify (through RFID tags) the cows for milk preparation. To achieve this smart milking, intelligent monitoring with sophisticated robots is required. The automated feeding is enabled by the systems like unmanned ground vehicles for feeding the cows every 4–5 hours. Feeding robots are filled with food based upon the cow's type. The health of the cows is also monitored to produce quality milk. Through smart dairy farming with advanced technology, the condition of the cows is also monitored to check whether the cow is pregnant.

8.3 IoT-based agriculture products

IoT in agriculture has created a profound advancement in technology to enhance monitoring, automation and accurate solutions. To meet the challenges, requirements and global standards, entrepreneurs are making a path for modern agriculture methods. In this section, agricultural products are described to understand the importance of IoT in agriculture. These products not only solve the real-time problems but also give an estimation of future parameters based on past data. The different IoT-based agricultural products and their usages are tabulated in Table 8.4.

Apart from the agricultural products mentioned in Table 8.4 many more devices exist like SKY-Lora that transfers the weather data to the connected sensor placed at

Table 8.4 Agriculture products and usage

Serial number	Product name	Pictorial	Usage
1.	AllMETEO		Accurate weather measurement (several parameters)
2.	Pycno (pycno.co, 2019)		System control and continuous weather data monitoring (data visualization, integrated soil sensor)
3.	GreenIQ (Ben nassi, 2018)		Garden-based smart sprinkler. Saves 50% water consumption
4.	CropX starter Kit (CropX Starter, 2019)		Uninterrupted soil temperature monitoring sensor with better accuracy features
5.	3D Crop Sensor Array (3D Crop Sensor, 2019)		Easily mountable and monitors carbon dioxide, humidity and temperature
6.	Arable Mark (arable.com, 2019)		Connects the global weather data under the respective field observations
7.	EC-1 controller (Growlink.com, 2019)		Feedback-based decision-making environmental condition monitoring system
8.	Growlink (Growlink.com, 2019)		Single microcontroller–based smart farming through IP network
9.	Smart element (smartelements.io)		Determines the wetness of a leaf
10.	Waspmote Plug		Reliable and accurate weather information

a distance of 600 m through Wi-Fi connection (get.pycno.co). Agriculture drones are more sophisticated for agriculture optimization, crop production improvement and crop monitoring through digital camera attached to the drone.

8.4 Privacy and security issues in smart farming

Smart farming includes the smart devices that are connected to the servers through a network. Smart farming increases the flexibility of agricultural operations through customer integration (Kamilaris *et al.*, 2016). From the numerous benefits of smart farming, there are some unavoidable risks. As smart farming is dependent on information systems for sustainable development, hackers and other competitors are lusting after the intellectual property (Deloitte, 2016). Digitization and growing technology are creating opportunities for hackers and terrorists to cyberattack and physically attack over prohibited data and places. It is a big problem that cannot be ignored (Olcott, 2016). So, privacy and cyber security awareness must be created to avoid the risks in smart farming or agriculture sector.

8.4.1 Privacy issues

In smart farming several sensor-based technologies are adopted over cloud through smart applications. The data provided by the sensors have to be healthy and should be protected. So, the privacy issues must be considered for a faithful work environment.

a. Data security: IoT sensors and devices collect enormous potential and dynamic data. Data leakage must be prevented to unauthorized persons.
    Example: Use of anti-jamming agriculture devices bypasses some several economic losses to the farmers. Smart farming includes artificial intelligence and "5G" communication that requires quick response from the users. So, the data are stored in the cloud. Privacy features must be provided to avoid data theft by third parties. A technology invention includes multi-disciplinary efforts in hardware and software. All the policies must be made against the occurrence of cyberattacks (Society, 2016).
b. Authorization: Smart farming includes devices (sensors) and autonomous vehicles (tractors, drones, etc.) that communicate with each other either through machine to machine or through cloud using message queue telemetry transport (http://mqtt.org), and some other protocols like constrained application protocols (http://coap.technology/) and IoT communication protocols. In agriculture, livestock (cattle) is a big source of income for farmers. In many cases, sensors are embedded in livestock for monitoring their health condition so that medicine can be injected remotely without doctor's intervention (Berckmans, 2006; Rosell-Polo *et al.*, 2015). While purchasing/selling the cattle if the sensors are not removed, important livestock data can be easily accessed. The equipment firmware must be updated over the air from authorized person(s) only. Multi-cloud and cross-cloud model must be used for firm updating and data transfer (Tang and Sandhu, 2013; Pustchi *et al.*, 2015). Proper authorization must be given to the farmer for making the right decisions for smart farming.

c. Secure communication: the most important aspect in privacy issues is providing secure communication through authentication of IoT devices. The public key infrastructure authentication mechanism is not a convenient solution for IoT devices because the IoT devices usually have less memory (storage), low power consumption and limited processing capability. Somewhat realistic solutions for smart farming would be secure and lightweight multi-factor protocols (Law *et al.*, 2013; Wazid *et al.*, 2018). Authentication of connecting device is done by the intermediary certifying authority (Fan *et al.*, 2020). For smart farming, the number of attempts made over lightweight cryptography is limited (Henriques and Vernekar, 2017).

d. Regulations: Smart farming raises different legal issues due to involvement of different stakeholders and bodies. So, different parties must come to an agreement and understand the license polices (Boghossian, 2018; Jahn, 2019). The farmers who invest in smart infrastructure fear data theft. In fact, the livestock and agriculture are merely regulated firms. Over the world various rules, regulations and authorities over selling and producing goods are there. Such amenability can be regulated by monitoring production cycle (https://jahnresearchgroup.webhosting. cals.wisc.edu; https://www.epa.gov).

8.4.2 Security threats

a. Denial of service (DOS): Smart farming is the interconnection of IoT devices through the web. The more reliable the connection, the less vulnerable it is to threats. However, when more devices, nodes and groups exist it results in DOS attacks. The DOS helps to block or stop working on its devices. One method to perform the DOS attack is by putting the data in a Faraday cage (Vacca, 2017). The cyber-attackers find the system vulnerabilities using DOS. These attacks cause material damage (Schreier, 2012). Sometimes DOS attacks are very difficult to control. So, preventing such attacks is the best thing.

b. Supply chain: The IoT and digital systems connect to a variety of organizational environments to make the supply chain more integrated and efficient. The main purpose of integration through the food supply chain is to create adequate job opportunities in rural areas. But, in agriculture as several products have a shorter lifetime the supply chain becomes uncertain. Optimizing the supply chain using IoT technology is a future challenge (Dougados *et al.*, 2013; Verdouw *et al.*, 2015). By creating the authentication mechanism through access control, security awareness and cryptographic process, the supply chain attacks can be prevented (e.g., Blockchain technology, http://bita.studio/).

c. Ransomware: It is a serious threat to individuals and businesses worldwide. The IoT and information and communication technology advancements in agriculture sector face this threat. The attackers use ransomware for encrypting the farmers' data to an unreadable format, so that they can demand a ransom to make it readable (Luo & Liao, 2009).

The global industries and individuals are looking to set up a stable foundation for smart farming and IoT applications to make them less vulnerable to cyber and security threats.

8.5 Applications of IoT in sustainable farming

The IoT technology plays a vital role in agriculture farming. The country's economy is dependent on agriculture (Abellanosa & Pava, 1987). Sustainability in agriculture signifies the agricultural resource management to meet the human requirements while conserving the natural resources and environmental quality. There are several issues in current farming practices like excessive use of fertilizers (Gomez & Thivant, 2015). In continuous agriculture too much water or too little water impacts soil texture and growth of crops (Anand et al., 2015). The conventional farming practices use synthetic fertilizers such as phosphorous and nitrogen compounds, whereas organic farming involves biodegradable fertilizers like animal manure which are less costly and not hazardous. Thus organic farming improves the sustainability (Ansari & Mahmood, 2017; Seufert et al., 2017).

8.5.1 Precision farming

Precision farming includes the farm management (adopting new techniques for improving the productivity using farm management system), which plays the main role in planning, making the decisions, processing to achieve the attributes of smart farming. It also includes micro-controllers, grain storage management modules and WSNs. The farms that are far away from the natural resources like lakes or rivers must utilize ground water resources. To provide the ground water resources, solar power supply has to be arranged. All such problems can be solved through precision farming. It is a new kind of farm management using information technology to collect precise information about soil, crop and weather conditions for optimal crop production (Pierce et al., 1997; Paucar et al., 2015; http://www.cema-agri.org/page/3-precision-crop-management. Precision farming includes greenhouse farming, indoor farming and vertical farming with sensors and satellite image technology.

8.5.2 Greenhouse farming

It is the best option to achieve precision farming targets. According to the geographical area, the temperature and climatic conditions are varied.

This greenhouse farming creates suitable temperature environment for the plants by the use of greenhouse technology. Rain, wind and pest protection can be achieved for best production (Cemek et al., 2006). The limitation of this farming is that it lacks pollination due to isolation from the insects (Malinowski, 2014).

8.5.3 Indoor vertical farming

The vertical farming focuses only on supplying nutrients and water directly to the roots. The physical conditions of crops can be monitored automatically through internet. It reduces the human efforts in the plants' pre- and post-growing period. The limitation of the indoor vertical farming is the number of crop type production.

8.5.4 Satellite imaging

It is also known as remote sensing where the crops can be monitored through satellites. The information about the fields is gathered by scanning the fields with a high-flying flight

Table 8.5 Mechanization of technology usage in precision farming

Country	Percentage of technology mechanization
Japan	99
South Korea	97
USA	95
Australia	90
China	91
Brazil	75
India	40–45

Source: Ramdinthara and Shanthi Bala (2019), Srisruthi et al. (2016), Hadas (2012).

or by satellites. It collects data about the crop conditions and soil to manage the crop properly as an optimal irrigation system (Sheffield and Morse-Mcnabb, 2014; Khirade & Patil, 2015; Zhang et al., 2016). The limitation of this farming is that it is very costly.

8.5.5 Comparative study

Countries around the world are implementing new technologies for higher production, better way of farming and conservation of the ecosystem. The comparative study of technology mechanization in India, USA, Japan, China, etc. is mentioned in Table 8.5. From the table, it can be seen that except for Brazil and India, remaining countries are maintaining 90% and above in the technology used for better production. These two countries are also trying to adopt the new technologies and machinery for meeting the requirements of the population. All these countries implemented and adopted sophisticated technologies for better production according to the population increasing rate.

8.5.6 Waste management

The IoT enables information technology integration and its application (Kim et al., 2017). The analytical framework for sensor-based IoT helps to achieve environmental sustainability (Bibri, 2018). The circular economy system enables to use energy and natural resources while minimizing waste through the principles of recycling, reducing and reusing (Tura et al., 2019). The circular economy loop includes the features of non-renewable and renewable as input followed by convergence, production, consumptions and finally residuals and emissions as output (Walmsley et al., 2019). In agriculture the waste can be prevented by estimating the future data through IoT devices. Providing the future market prices of the agriculture product helps the farmers sell the goods at the right time. The features are described as follows:

Minimization and supply: Data collection sensors, waste amount detection, waste reduction.
Distribution and collection: Collecting each type of waste frequently, treatment of the plant by detecting diseases of the plants at early stage and choosing the right treatment.
Technology selection: Safety sensors, parameter control by selecting the proper technology.

8.5.7 Smart drone farming

The drones with IoT finds several applications in farming and smart cities. A drone-enabled intelligent and smart city can be implemented for future generations (Menouar *et al.*, 2017; Alam *et al.*, 2018). An IoT-based delivery service through drone architecture is described (Motlagh *et al.*, 2016). IoT-based services for large coverage can be implemented for face recognition and crowd surveillance (Motlagh *et al.*, 2017). A drone that is equipped with camera, different sensors and pesticide-carrying capability will perform better in spraying pesticide and field monitoring and thus reduce the labor cost also. The drone is not only confined to farming but covers the applications in different categories such as object detection and tracking, surveillance, data collection (Wi-Fi, 5G, RFID), path planning and navigation, pollution estimation, traffic monitoring, security and emergency services.

8.6 Conclusion

The IoT has potential in majority of fields. The advancement in technology and availability of agriculture sensors connected through different wireless protocols open the doors for farmers and consumers to improve the production. However, there are some security and privacy challenges to be handled properly. In contrast to conventional farming, IoT-based smart farming provides accurate and feasible features in crop production, waste management and cost-effective solutions. All over the world the scientists and researchers are doing research on services that can be provided through IoT technology. The governments of different countries are funding IoT-based research centers to replace conventional farming with smart farming.

Bibliography

Abellanosa, A.L., and H.M. Pava. (1987): *Introduction to Crop Science: Central Mindanao University Publications Office*, Musuan, Bukidnon: Publications Office.

Alam, Muhammad, Davide Moroni, Gabriele Pieri, Marco Tampucci, Miguel Gomes, José Fonseca, Joaquim Ferreira, and Giuseppe Riccardo Leone. (2018): Real-Time Smart Parking Systems Integration in Distributed ITS for Smart Cities. *Journal of Advanced Transportation*. [Online], 1–13. https://doi.org/10.1155/2018/1485652.

Anand, Koushik, Jayakumar, Mohana Muthu, and Sridhar Amirneni. (2015): Automatic Drip Irrigation System Using Fuzzy Logic and Mobile Technology, *IEEE Technological Innovation in ICT for Agriculture and Rural Development, TIAR 2015*, 10–12 July 2015, Chennai, India. pp. 54–58.

Ansari, Rizwan and Mahmood, Irshad. (2017): Optimization of Organic and Bio-Organic Fertilizers on Soil Properties and Growth of Pigeon Pea. *Scientia Horticulturae*. 226, 1–9. doi: 10.1016/j.scienta.2017.07.033.

Berckmans. (2006): *Automatic On-Line Monitoring of Animals by Precision Livestock Farming Livestock Production Society*. Wageningen Academic Publishers, Wageningen, Netherlands.

Bhanu, Balaji, Raghava Rao, Ramesh, and Mohammed Ali Hussain. (2014): Agriculture Field Monitoring and Analysis Using Wireless Sensor Networks for Improving Crop Production. *Eleventh International Conference on Wireless and Optical Communications Networks, WOCN 2014*, 11–13 September 2014, Vijayawada, India, pp. 1–7.

Bibri, Simon Elias. (2018): The IoT for Smart Sustainable Cities of the Future: An Analytical Framework for Sensor-Based Big Data Applications for Environmental Sustainability.

Sustainable Cities and Society. [Online] 38, 230–253. Available from: https://doi.org/10.1016/j. scs..2017.12.034.

Boghossian. (2018): Threats to Precision Agriculture: Technical. Report, U. S. Department of Homeland Security. DOI: 10.13140/RG.2.2.20693.37600.

Cemek, Bilal, Yusuf Demir, Sezgin Uzun, and Vedat Ceyhan. (2016): The Effects of Different Greenhouse Covering Materials on Energy Requirement, Growth and Yield of Aubergine. *Energy.* [Online] 31 (12), 1780–1788. Available from: https://doi.org/10.1016/j.energy.2005. 08.004.

Dan, Liu, Cao Xin, Huang Chongwei, and Ji Liangliang. (2015): Intelligent Agriculture Green-house Environment Monitoring System based on IOT Technology. *2015 International Conference on Intelligent Transportation, Big Data and Smart City*, 19–20 December 2015, Halong Bay, Vietnam, pp. 487–490.

Deloitte. (2016): Hacktivism: A Defender's Playbook: Techical Report, Vigilant Threat Studies, Deloitte US.

Dougados, Ghioldi, Doesburg, and Subrahmanya. (2013): Digital Technologies Offer a Shot in the Arm for Traditional Supply Chains cap Gemini. Available from: https://www.capgemini. com/wp-content/uploads/2017/07/supply_chain_paper_13-12_cc.pdf.

DTAC Debuts the First IoT Based Agricultural Solution. (2019): https://www.telenor. com/dtac-debutsthe-_rst-iot-based-agricultural-solution/.

Duy, Nguyen Tang Kha, Nguyen Dinh Tu, Tra Hoang Son, and Luong Hong Duy Khanh. (2015): Automated Monitoring and Control System for Shrimp Farms Based on Embedded System and Wireless Sensor Network. *2015 IEEE International Conference on Electrical, Computer and Communication Technologies, ICECCT 2015*, 5–7 March 2015, Coimbatore, India, pp. 1–5.

Fan, Xinxin, Ling Liu, Rui Zhang, Quanliang Jing, and Jingping Bi. (2020): Decentralized Trust Management. *ACM Computers Survey.* https://doi.org/10.1145/3362168.

Farm Institute. (2019): IoT in Agriculture: How Is It Evolving. http://www.farminstitute.org. au/LiteratureRetrieve.aspx?ID=157672.

Food and Agriculture. (2019): https://www.nectec.or.th/en/research/.

Gomez, Ilka and Lisa Thivant. (2015): Training Manual for Organic Agriculture, Food and Agriculture Organization of the United Nation (FAO)

Hadas (2012): Ancient Agricultural Irrigation Systems in the Oasis of Ein Gedi, Dead Sea, Israel. *Journal of Arid Environments.* [Online] 86, 75–81. Available from: https://doi.org/ 10.1016/j.jaridenv.2011.08.015.

Haule, Joseph, and Kisangiri Michael. (2014): Deployment of Wireless Sensor Networks (WSN) in Automated Irrigation Management and Scheduling Systems: a Review. *Proceedings of the 2nd Pan African International Conference on Science, Computing and Telecommunications, PACT 2014*, 14–18 July 2014, Arusha, Tanzania, pp. 86–91.

Henriques, Michelle , and Nagaraj Vernekar. (2017): Iciot 2017, Using Symmetric and Asym-metric Cryptography to Secure Communication between Devices in IoT. *2017 International Conference on IoT and Application, ICIOT 2017*, 19–20 May 2017, Nagapattinam, India, pp. 1–4.

Jahn (2019): Cyber Risk and Security Implications in Smart Agriculture and Food Systems. https://jahnresearchgroup.webhosting.cals.wisc.edu.

Kamilaris, Andreas, Feng Gao, Francesc X. Prenafeta-Boldu, and Muhammad Intizar Ali. (2016): Agri-IoT: A Semantic Framework for Internet of Things-Enabled Smart Farm-ing Applications. *2016 IEEE 3rd World Forum on Internet of Things, WF-IoT, 2016*, 12–14 December 2016, Reston, VA, USA, pp. 442–447.

Khirade, Sachin, and Patil. (2015): Plant Disease Detection Using Image Processing. *2015 International Conference on Computing Communication Control and Automation, ICCUBEA 2015*, 26–27 February 2015, Pune, India, pp. 768–771.

Kim Tai-hoom, Carlos Ramos, and Sabah Mohammed. (2017): Smart City and IoT. *Future Generation Computer System*. 76, 159–162.

Law Yee Wei, Marimuthu Palaniswami, Gina. Kounga, and Anthony Lo. (2013): WAKE: Key Management Scheme for Wide-Area Measurement Systems in Smart Grid. *IEEE Communications Magazine*. 51 (1), 34–41. https://doi.org/10.1109/mcom.2013.6400436.

Luo X. and Q Liao. (2009): Ransomware: A New Cycber Hijacking Threat to Enterprises. [Online] *Handbook of Research on Information Security and Assurance*. IGI Global Available from: http://services.igi-global.com/resolvedoi/resolve.aspx?doi=10.4018/978-1-59904-855-0.ch001.

Malinowski. (2014). Pros and Cons of Greenhouse Growing, MidAtlantic Farm.

Menouar, Hamid, Ismail Guvenc, Kemal Akkaya, Selcuk Uluagac, Abdullah Kadri, and Adem Tuncer. (2017): UAV-Enabled Intelligent Transportation Systems for the Smart City: Applications and Challenges. *IEEE Communications Magazine*. 55(3), 22–28. https://doi.org/10.1109/mcom.2017.1600238cm.

Mooney, Peter. (2017): Open Source Farming the next Agricultural Revolution, [Lecture] Annebrook House Hotel, 24th January.

Motlagh, Naser Hossein, Miloud Bagaa, and Tarik Taleb. (2017): UAV-based IoT Platform: A Crowd Surveillance Use Case. *IEEE Communications Magazine*. 55(2), 128–134. https://doi.org/10.1109/mcom.2017.1600587cm.

Motlagh, Naser Hossein, Tarik Taleb, and Osama Arouk. (2016): Low-Altitude Unmanned Aerial Vehicles-Based Internet of Things Services: Comprehensive Survey and Future Perspectives. *IEEE Internet of Things Journal*. 3(6), 899–922.

Ojha, Tamoghna, Sudip Misra, and Narendra Singh Raghuwanshi (2015): Wireless Sensor Networks for Agriculture: The State-of-the-Art in Practice and Future Challenges. *Computers and Electronics in Agriculture*. [Online] 118, 66–84. doi:10.1016/j.compag.2015.08.011.

Olcott. (2016): Input to the Commission on Enhancing National Cybersecurity: The Impact of Security Ratings on National Cyber Security: Tech. Rep.

Paucar, L. G., Diaz, A. R., Viani, F., Robol, F., Polo, A., and Massa, A. (2015): Decision Support for Smart Irrigation by Means of Wireless Distributed Sensors. *Paper resented at Mediterranean Microwave Symposium*, https://doi.org/10.1109/MMS.2015.7375469.

Pierce, Sadler. (1997): *The State of Site Specific Management for Agriculture*. ASA Publications, CSSA e SSSA, Madison, WI, USA.

Pustchi, Navid, Ram Krishnan, and Ravi Sandhu. (2015): Authorization Federation in IaaS Multi Cloud. *Proceedings of the 3rd International Workshop on Security in Cloud Computing - SCC 15*, 2015. https://doi.org/10.1145/2732516.2732523.

Ramdinthara, Immanuel Zion, and Shanthi Bala. (2019): A Comparative Study of IoT Technology in Precision Agriculture. *IEEE International Conference on System, Computation, Automation and Networking, ICSCAN 2019*, 29–30 March 2019, Padichherry, India, pp. 1–5.

Rosell-Polo, Joan R., Fernando Auat Cheein, Eduard Gregorio, Dionisio Andújar, Lluís Puigdomènech, Joan Masip, and Alexandre Escolà (2015): Advances in Structured Light Sensors Applications in Precision Agriculture and Livestock Farming. *Advances in Agronomy*, 71–112. https://doi.org/10.1016/bs.agron.2015.05.002.

Sales, Nelson, Orlando Remedios, and Artur Arsenio. (2015): Wireless Sensor and Actuator System for Smart Irrigation on the Cloud. *2015 IEEE 2nd World Forum on Internet of Things (WF-IoT), 2015*, 14–16 December 2015, Milan, Italy, pp. 693–698.

Schreier, F. (2012): On Cyberwarfare: DCAF Horizon Working Paper, No. 7, pp. 1–133. Available from: http://www.dcaf.ch/Publications/On-Cyberwarfare.

Seufert, Verena, Navin Ramankutty, and Tabea Mayerhofer (2017): What Is This Thing Called Organic? – How Organic Farming Is Codified in Regulations. *Food Policy*. [Online] 68, 10–20. https://doi.org/10.1016/j.foodpol.2016.12.009.

Sheffield, K., and E. Morse-Mcnabb. (2014): Using Satellite Imagery to Asses Trends in Soil and Crop Productivity across Landscapes. *IOP Conference Series: Earth and Environmental Science, IOPCS 2014*, 24–27 March 2014, Bendigo, Victoria, Australia, pp. 1–11.

Sivasankari, A., and S. Gandhimathi (2014): Wireless Sensor Based Crop Monitoring System for Agriculture Using Wi-Fi Network Dissertation. *International Journal of Computer Science and Information Technology*. [Online] Vol. 2, issue 3, pp. 293-303.

Society. (2016): Cybersecurity Opportunities, Threats Challenges: Tech. Rep.

Srisruthi, S., N. Swarna, G. M. Susmitha Ros, and Edna Elizabeth. (2016): Sustainable Agriculture Using Eco-Friendly and Energy Efficient Sensor Technology. *IEEE International Conference on Recent Trends in Electronics, Information & Communication Technology, RTEICT*, 20–21 May 2016, Bangalore, India, pp. 1442–1446.

Tang, Bo, and Ravi Sandhu. (2013): Cross-Tenant Trust Models in Cloud Computing. *IEEE 14th International Conference on Information Reuse & Integration, IRI 2013*, 14–16 August 2013, San Francisco, CA, USA, pp. 129–136.

Tura, Nina, Jyri Hanski, Tuomas Ahola, Matias Ståhle, Sini Piiparinen, and Pasi Valkokari. (2019): Unlocking Circular Business: A Framework of Barriers and Drivers. *Journal of Cleaner Production*. [Online] 212, 90–98. https://doi.org/10.1016/j.jclepro.2018.11.202.

United States, Environment Protection Agency. (2019): Available from: https://www.epa.gov/agriculture/laws-and-regulations-apply-your-agricultural-operationfarm-activity

Vacca. (2017): *Computer and Information Security Handbook*: 3rd Edition. Amsterdam: Elsevier.

Verdouw Cor, Adrie Beulens, Hajo Reijers, and Jack Van der Vorst. (2015): A Control Model for Object Virtualization in Supply Chain Management. *Computers in Industry* 68, 116–131. https://doi.org/10.1016/j.compind.2014.12.011.

Vijayakumar., and Nelson Rosario. (2011): Preliminary Design for Crop Monitoring Involving Water and Fertilizer Conservation Using Wireless Sensor Networks. *2011, IEEE 3rd International Conference on Communication Software and Networks*, 27–29 May 2011, Xi'an, China, pp. 662–666.

Walmsley, Timothy Gordon, Benjamin H.Y. Ong, Jiří Jaromír Klemeš, Raymond R. Tan, and Petar Sabev Varbanov. (2019): Circular Integration of Processes, Industries, and Economies. *Renewable and Sustainable Energy Reviews*, 107, 507–515. https://doi.org/10.1016/j.rser.2019.03.039.

Wazid, Mohammad, Ashok Kumar Das, Vanga Odelu, Neeraj Kumar, Mauro Conti, and Minho Jo. (2018): Design of Secure User Authenticated Key Management Protocol for Generic IoT Networks. *IEEE Internet of Things Journal*. [Online] 5(1), 269–282. https://doi.org/10.1109/jiot.2017.2780232.

Zhang, Lin, Min Yuan, Deyi Tai, Jian Ding, Xiaowei Xu, Xiang Zhen, and Yuanyuan Zhang. (2010): Design and Implementation of Granary Monitoring System Based on Wireless Sensor Network Node. *2010 International Conference on Measuring Technology and Mechatronics Automation*, 13–14 March 2010, Changsha City, China, pp. 950–953.

Zhang, Wen, Zhiyuan Zhang, and Zhen Huang. (2016): Auto-Extraction Method of Farmland Irrigation and Drainage System based on Domestic High-Resolution Satellite Images. *2016 Fifth International Conference on Agro-Geoinformatics (Agro-Geoinformatics), Agro-Geoinformatics 2016*, 18–20 July 2016, Tianjin, China, pp. 1–6.

Chapter 9

Smart farming

Crop models and decision support systems using IoT

Kanwaljeet Singh and Amandeep Kaur

LOVELY PROFESSIONAL UNIVERSITY

9.1 Introduction

Agriculture is one of the oldest and predominant fields in every country's economic growth that acts as its backbone. The origin of the word agriculture is from the Latin word "ager", which means land, and "culture", which means cultivation. Like India in some other countries also the largest population income depends on farming practices. This sector is also considered as one of the biggest sources for employment and income generation. As per a survey report, in India nearly 64% of the total population is engaged in agriculture, mainly people from rural areas. Agriculture is the only source of food generation like fruits and vegetables. As compared with ancient times, now many tools and fast techniques are available in farming sector. Besides these latest techniques, the environmental factor plays an important role. In India, farmers are facing many problems concerned with agriculture like increasing demand of food products and weather-changing patterns that mostly affect the growth of crops these days. As per Unified National Fine (UNF) Food and Agriculture Organization, to fulfill the food requirements of increasing population about 70% more food has to be produced and that it was produced by 2006, so the old fashioned agriculture practices have moved towards smart farming with the help of agriculture companies and awareness of farmers.

Moreover, crop selection highly depends upon natural factors like temperature, moisture, sun light, humidity, etc. Also soil conditions are important like soil type, fertility and pH value. Adequate water is also required for irrigation purposes. In our country, water supply to crops depends upon monsoon which is not sufficient for irrigation purposes. How much watering is needed also varies from soil category and moisture content of that soil.

To fulfill all these challenges, companies are using concept of internet of things (IoT), which has paved the way towards smart and efficient farming methods. This IoT field of technology plays a quintessential role in boosting up crop yield rate, increasing global market connectivity by smart observing activities, enhancing weather monitoring to nurture different plants and making weather predictions, etc., with minimum human involvement within less time, thus making systems more cost efficient. By smart IoT devices, farmers can observe their fields and can get information easily from worldwide locations. In agriculture field IoT is gaining more attention due to the following key points:

1. Global connectivity is possible by IoT operated devices
2. Farming activities can be monitored with less human efforts
3. Faster access
4. Time efficiency

During the last few decades, the most critical issue in traditional farming methods has been climate change. It includes heat waves, rainfall patterns, storms, droughts, etc. All these factors play a predominant role in reduced productivity. These factors also affect the life cycle of plant growth. So, to counter these issues and challenges IoT is used extensively in agriculture field. The world food organizations are expecting agriculture market to reach 18.45 billion dollars in 2020. So, IoT devices are of great help in enhancing productivity in agriculture field as they are used to track soil quality, temperature of fields, humidity, safety from animals etc. It also helps in monitoring the live stocks and quality of crops. To accomplish these activities numerous sensors are used to provide real-time data to farmers related to rainfall, soil nutrition and pest infections.

Main benefits of IoT in farming are as follows:

1. Efficient water management without any wastage using sensors
2. Monitoring of soil conditions before growing plants
3. Reduces manpower and cuts down the labor cost
4. Real-time monitoring for proper growth of crops
5. Identification of soil moisture and pH level
6. Identification of plant diseases using sensors and Radio-frequency identification (RFID) chips. This information can be shared with scientists and action can be taken to protect the crops.
7. Sale of crop will be boosted up without any geographical restrictions

9.2 Some applications of IoT in agriculture

Water management
Soil monitoring
Routine operations
Monitoring and control systems
Smart irrigation system
Agri-produce and agri-resource management

In many parts of the country, for watering of crops different irrigation methods are used like tube wells, canals and by pumping ground water with the help of heavy motors that operate on electricity. Ground water pumping with motor depends upon electricity availability during day and night times which leads to high electricity consumption. Sometimes due to lack of continuous monitoring by farmers, fields are over-watered which leads to wastage of resources. So, IoT technology is used to track this practice in more effective way by using moisture sensors. This article presents how moisture is detected with the help of sensors, Arduino IDE (integrated development board) and Blynk app (Figure 9.1).

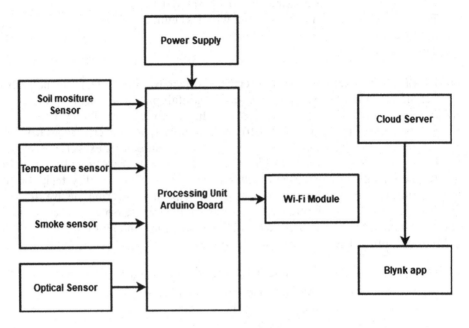

Figure 9.1 Block diagram of smart agriculture system.

9.3 Description of tools

The following tools/components were used to build this project:

Arduino UNO board and Arduino IDE
Buzzer
Smoke sensor or MQ-2
Soil moisture and temperature sensor
Optical sensor
Analog-to-digital converter (ADC)
Blynk app
Relays
ESP8266

9.3.1 Arduino

Arduino is an organization which designs software and develops open source hardware and software based on microcontroller boards. Electronics projects are built using this open source board. Different types of Arduino boards are available like UNO, Lily pad, Nano, Mega etc. Arduino UNO is based on microcontroller ATMga328. It has 14 digital input/output pins, out of which 6 pins are PWM pins, 6 analog input pins, a USB connector to give 5 V DC power supply, quartz crystal with 16 MHz frequency, reset button and a power jack. It has 32 Kbytes of chip memory, out of which

0.5 kb is used by boot loader. It is a programmable circuit board, and programs can be written and uploaded on board using Arduino IDE.

An Uno board is the first version in series of USB Arduino boards. Arduino boards can read inputs from analog and digital sensors, can take input from switch and convert it into output signal to control DC motor, to blink LEDs and display at output.

Arduino UNO is easy to use as compared with microcontrollers and microprocessors and used by beginners to design different projects and applications based on it. It is more flexible and can be run on different platforms like Windows, Mac and Linux. In every field like medical, agriculture, transportation and home automation, Arduino-based projects can be designed with less complexity and limited cost.

As compared with microcontrollers, Arduino boards simplify the task to work on microcontrollers using Arduino IDE with different library functions. These boards become point of attraction due to following advantages:

Inexpensive: Arduino boards are cheap in cost in comparison with other microcontrollers like PIC. Moreover, these boards can be assembled on breadboard.

Simple and clear programming: For beginners, to learn and create projects Arduino is very simple and easy-to-use board. Programming can be done using Arduino IDE and plenty of examples are available in IDE help. Advanced programming can also be done to design complex projects by adding libraries in it (Table 9.1).

9.3.1.1 Arduino UNO description

Power: Arduino UNO board can be powered by external power supply using USB connection (Figure 9.2). It works on 5 V DC power supply, so board can be either connected with laptop through USB cable, or adapter can be used which converts AC into DC power supply. Board can also operate on 6–20 V power supply. Some of the power supply pins used on board are explained below:

Vin: If board is powered through external power source, this pin can be used. Supply can be given through this pin or can be accessed also when power is given by USB port.

Table 9.1 Arduino UNO technical specifications

Microcontroller	ATmega328
Clock speed	16 MHz
Operating voltage	5 V
Maximum supply voltage (not recommended)	20 V
Supply voltage (recommended)	7–12 V
Analog input pins	6
Digital input/output pins	14
DC current per input/output pin	40 mA
DC current in 3.3 V pin	50 mA
SRAM	2 kb
EEPROM	1 kb
Flash memory	32 kb of which 0.5 kb used by boot loader

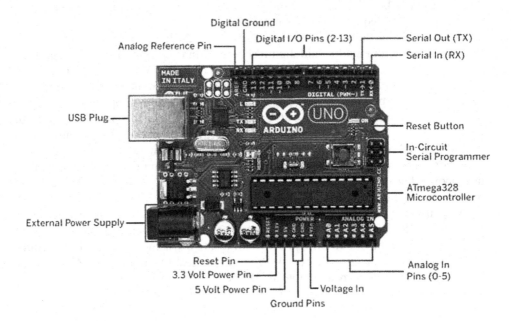

Figure 9.2 Arduino UNO board description.

5 V: Power supply on this pin is regulated by 5 V DC power supply that is obtained from onboard regulator. Input power to voltage regulator is supplied through DC power jack or USB connector.

GND (Ground): This pin can be used to make GND connections.

IOREF: The main use of this pin is to supply reference voltage. This pin mainly works with 3.3 V DC power supply.

Memory: In Arduino UNO, ATMega microcontroller has 32 Kbytes of on-chip flash memory that is used by bootloader to upload Arduino IDE sketch. Moreover, 2 Kbytes of SRAM is used for runtime data usage and it also has 1 Kbyte of EEPROM (electrically erasable programmable read-only memory) memory.

Input and output pins: Arduino has two types of pins on board: digital pins and analog pins. Pins from 0 to 13 are used as digital input and output pins to receive and send digital data and these pins can be configured as in I/O mode using pinmode() command. These pins can be marked as LOW and HIGH using digital functions like digitalWrite() and can be read using digitalRead() functions.

9.3.2 Buzzer

A buzzer is a small and very efficient device that is used to generate sound signal, so it is called beeper also known as audio signal generating device. It can be a mechanical, electro-mechanical or piezoelectric device. Mainly it is used for alarms, timers and to check inputs initiated by users like mouse click or keystroke. It is a two-pin device and

Figure 9.3 Buzzer.

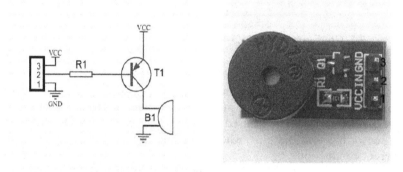

Figure 9.4 Schematic diagram of buzzer and its pin configuration.

can be easily integrated on board. It consists of electronic transducer and DC power supply (Figures 9.3 and 9.4).

Buzzers are mainly used in following applications:
Novelty uses
Judging panels
Educational purposes
Annunciator panels
Electronic metronomes
Pin configuration
VCC (Common Collector Voltage)
Input
Ground

9.4 MQ-2 semiconductor sensor for combustible gas

MQ-2, the grove gas sensor is mainly used to detect leakage of gas from home and industry. It can detect different gases like H_2, CH_4, LPG, CO, alcohol, smoke and propane. It is very fast in taking actions because it has very high sensitivity and very fast response time. It is a metal oxide semiconductor gas sensor and also known as

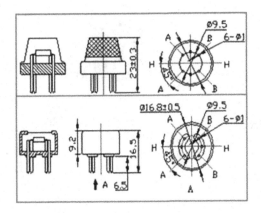

Figure 9.5 Configuration of MQ-2.

chemiresistor because gas detection is done by measuring change in resistance when gas comes in contact with sensor material. To detect different concentrations of gas, voltage divider network can be used. This sensor works on 5 V DC proper supply. The MQ-2 sensor has total six pins, where four pins are used for signal fetching and two for current conduction.

Sensor consists of two layers of stainless steel known as anti-explosion mesh network. Its role is to ensure that heater element which is inside the sensor does not cause any explosion as sensing process of flammable gases is going on (Figure 9.5).

In air at high temperature, when tin dioxide is heated, oxygen content is absorbed from the surface. Donor electrons are attracted towards oxygen which is absorbed from surface of sensing material in clean air and prevents the flow of electric current.

Applications

Domestic gas leakage detector
Industrial combustible gas detector
Portable gas detector
Air-quality check
Gas leakage alarm
To maintain environment standards in hospitals

Some conditions must be avoided:

9.4.1 Water condensation

Under indoor conditions, slight water condensation will affect sensor performance lightly. However, if water condensation is on sensor surface for a certain period of time, sensor's sensitivity will be decreased.

9.4.2 High gas concentration

No matter whether the sensor is electrified or not, if placed in high gas concentration for a long time some sensor characteristics will be affected.

9.4.3 Long time storage

The sensor resistance produces a reversible drift if it is stored for a long time without electricity; this drift is related to storage conditions. Sensors should be stored in air-proof condition without silicone gel bag with clean air. When any of the sensors are stored for a long time without electricity, it requires a long aging time for stabilizing before using.

9.4.4 Long time exposure to adverse environment

No matter whether the sensors are electrified or not, if exposed to adverse environment for a long time, such as high humidity, high temperature or high pollution, it will affect the sensor performance badly.

9.4.5 Vibration

Continual vibration will result in sensors' down-lead response then rupture. In transportation or assembling line, pneumatic screwdriver/ultrasonic welding machine can lead to this vibration.

9.4.6 Concussion

If sensors meet strong concussion, their lead wire may get disconnected.

9.4.7 Usage

Sensor handling can be done either manually or using wave crest welding. To use wave crest welding, following points should be kept in mind.

Soldering flux: Rosin soldering flux contains least chlorine
Speed: 1–2 m/min
Warm-up temperature: $100\pm20°C$
Welding temperature: $250\pm10°C$ times pass wave crest welding machine

If these conditions are not taken care of, then sensitivity of the sensor will be reduced.

9.5 Soil moisture sensor

Soil moisture sensor, as clear from its name, is used to detect the moisture level of soil. It detects presence of water content in soil depending upon checking the resistance or dielectric constant of soil. The measured value can be calibrated and can be affected

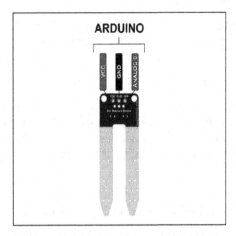

Figure 9.6 Soil moisture sensor.

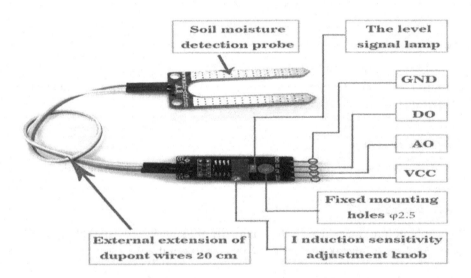

Figure 9.7 Soil moisture sensor pins.

by different parameters like type of soil, environmental temperature, conductivity of soil etc. These sensors are used by farmers or gardeners to check water volume of soil.

Soil moisture sensor is an analog machine and mainly works on capacitance principle to detect the water content in soil, for example, it records the dielectric permittivity of soil which is a function of water content. Simply it is inserted it in the soil and it shows the output of moisture level in the range between 0 and 1023 (Figures 9.6 and 9.7).

Applications

Agriculture

Landscape irrigation
Gardening

9.6 Relay

Relays are generally type of switches used to operate electrical ON and OFF connections electronically or electro-mechanically. Relays are used to control a circuit by establishing and closing contacts with other circuits. When relay is not powered, it is in normally open (NO) condition, and when there is close contact it is normally closed (NC) again as it is not energized. After applying the electrical current, the switches will change their state.

Relays are mainly used to make flow of smaller currents in control circuits. They cannot control high-power consuming devices except small motors and solenoids that control low amperes. By using amplifying concept, relays can also control large amperes and voltages. Protection feature also exists in relays including over-currents, overloads, reverse currents and undercurrents to prevent equipment damage. Relays are also used to switch the starting coils, heating elements and audible alarms (Figure 9.8).

In Figure 9.9, relay schematic diagram is represented. As seen from the diagram, the contacts at the top are NC (normally closed). When flow of current occurs through this contact, magnetic field is created which makes the switch closed. Mainly a spring is used to pull the switch to open position when power is removed from the energized coil.

There are different configurations of relays like single pole single throw (SPST) and single pole double throw (SPDT). SPST has two contacts, but SPDT has three contacts.

Figure 9.8 Relay.

Figure 9.9 Schematic representation of relay.

These contacts are usually labeled as COM (common), NO (normally open) and NC (normally closed). When no power is applied, normally closed contact should be connected to common terminal, and similarly normally open contact should remain open when there is no power supply given to coil.

In another way, when coil is powered up, common terminal is connected to normally open and normally closed terminal is left open. Double and single pole versions are almost same, except for two switches that open and close together.

9.7 Blynk application

Blynk app is used in IoT to design smart IoT devices easily and quickly. Data of sensors can be read, stored and visualized and as per that hardware can be controlled from remote locations. It works with both iOS and Android to control boards like Arduino, Node MCU and Raspberry Pi and links the data over the internet. Data can be plotted in graphical form also.

Blynk has Blynk server and library. Blynk offers secure and centralized cloud-based services through its server to communicate between devices. This server is available as open source. One can make its own server and be more secure. The most important feature is Blynk library that makes it more flexible. Hardware is connected through Blynk library and helps to run Blynk app. Many devices like Arduino, ESP8266 and Raspberry Pi are included in its library and help to connect with hardware through different methods of communication like Bluetooth, USB, Wi-Fi and GSM.

9.7.1 How Blynk works

Blynk works using three components:

Blynk app: Allows user to create their own interfaces for their projects using different widgets that users provide.

Blynk server: Used to provide communication between the hardware circuit and smartphones. By using Blynk cloud, users can run their own personal Blynk server locally. It is an open source and thousands of devices can be connected.

Blynk libraries: Used to run incoming and outgoing commands for all types of hardware platforms to enable communication with server (Figure 9.10).

9.7.2 Features

- Similar API (application programming interface) and UI (user interface) for all supported hardware and devices
- Connection to the cloud using:
- Wi-Fi
- Bluetooth and BLE (Bluetooth Low Energy)
- Ethernet
- USB (serial)
- GSM (Global System for Mobile Communications)
- Set of easy-to-use widgets

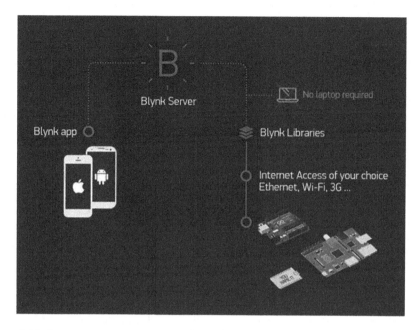

Figure 9.10 Components of Blynk app.

- Direct pin manipulation with no code writing
- Easy to integrate and add new functionality using virtual pins
- History data monitoring via History Graph widget
- Device-to-device communication using Bridge widget
- Sending emails, tweets, push notifications, etc.

9.7.3 Requirements for Blynk

Hardware: To work on Blynk platform, development kits are needed like Arduino, Node MCU, Raspberry Pi. These development kits should be connected with internet. For example, Arduino UNO board needs additional Wi-Fi module to communicate over the internet, but others are already Wi-Fi enabled like ESP8266, Raspberry Pi with Wi-Fi dongle, Particle Photon or SparkFun Blynk Board. It can be connected to laptop and desktop through USB.

A Smartphone: The Blynk app is a well-designed interface builder. It works on both iOS and Android (Figure 9.11).

One of the major benefits of ADC converter is its high data acquisition rate even at multiplexed inputs. With the invention of a wide variety of ADC integrated circuits, data acquisition from various sensors has become more accurate and faster. High-performance ADCs possess improved dynamic characteristics such as measurement repeatability, low power consumption, precise throughput, high linearity, excellent signal-to-noise ratio and so on.

These converters sample the analog signal on each falling or rising edge of sample clock. In each cycle, the ADC takes the analog signal, measures it and then

Figure 9.11 Snapshot of Blynk app.

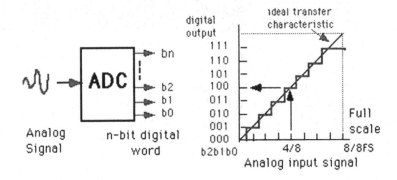

Figure 9.12 ADC conversion process.

converts it into a digital value. The ADC converts the output data into a series of digital values by approximating the signal with fixed precision. In ADCs, two factors determine the accuracy of the digital value that has been captured by the original analog signal. These are quantization level or bit rate and sampling rate. Figure 9.12 depicts how analog to digital conversion takes place. Bit rate decides the resolution of digitized output and in Figure 9.12 shows how a 3-bit ADC is used for converting analog signal.

The following tools/components were used to build this project:

- Arduino UNO and its IDE
- Buzzer
- Smoke sensor or MQ-2
- Soil moisture sensor
- ADC

- Blynk app
- Relays

We have used Arduino development kit to prepare the proposed system. The smoke sensor and the buzzer were powered and controlled by the Arduino micro-controller. The smoke sensor reacts to certain elements (chemicals) and triggers the buzzer, thereby alarming the farmer of supposed fire in the field.

The next half is done using ESP8266 Wi-Fi module, ADC, soil moisture sensor, relays, pump (represented by a bulb) and Blynk application.

The soil moisture sensor continuously measures the amount of moisture present in the soil. The ADC converts the analog signal of the sensor into a digital signal. The digital signal is then transferred to the Node MCU. The Node MCU is connected wirelessly to the Blynk app and its server system. When the moisture sensor detects that moisture content is less than the threshold, the Node MCU triggers the Blynk app and sends a notification to the app/device. The relay that is used to control the pump then turns off, allowing the pump to irrigate the fields.

When the soil moisture levels are restored back to the normal levels (detected using the soil moisture sensor), the relay is then triggered back by the Blynk app and Node MCU and the motor pump is turned off automatically.

9.8 Result and conclusion

The result of this project was a cost-efficient and easy-to-use system that can have applications in real world. It will help the farmers in drought-affected areas the most while conserving water and food grains.

On a personal level, the project was an enriching experience. It allowed us to have a hands-on experience on practical projects that can have deep and meaningful results for the society. The challenges were great but still we managed to overcome them and were able to deliver the given project in the stipulated time frame.

9.8.1 Scope

The scope for the project is enormous for a farming-oriented country like India. Advancements can be done with the help of additional sensors/software like pH scale detector, drone monitoring system, image processing and artificial intelligence.

Bibliography

Krishna, K. Lokesh, Omayo Silver, Wasswa Fahad Malende, and K. Anuradha. "Internet of Things application for implementation of smart agriculture system." In *2017 International Conference on I-SMAC (IoT in Social, Mobile, Analytics and Cloud) (I-SMAC)*, pp. 54–59. IEEE, 2017.

Mandula, Kumar, Ramu Parupalli, A. S. Murty, E. Magesh, and Rutul Lunagariya. "Mobile based home automation using Internet of Things (IoT)." In *2015 International Conference on Control, Instrumentation, Communication and Computational Technologies (ICCICCT)*, pp. 340–343. IEEE, 2015.

Patil, Gokul L., Prashant S. Gawande, and Rohit Vilasrao Bag. "Smart Agriculture System based on IoT and Its Social Impact." *International Journal of Computer Applications* 176, no. 1 (2017): 975–8887.

Patil, K. A., and N. R. Kale. "A model for smart agriculture using IoT." In *2016 International Conference on Global Trends in Signal Processing, Information Computing and Communication (ICGTSPICC)*, pp. 543–545. IEEE, 2016.

Rajalakshmi, P., and S. Devi Mahalakshmi. "IOT based crop-field monitoring and irrigation automation." In *2016 10th International Conference on Intelligent Systems and Control (ISCO)*, pp. 1–6. IEEE, 2016.

Suma, N., Sandra Rhea Samson, S. Saranya, G. Shanmugapriya, and R. Subhashri. "IOT based Smart Agriculture Monitoring System." *International Journal on Recent and Innovation Trends in Computing and Communication* 5, no. 2 (2017): 177–181.

Sushanth, G., and S. Sujatha. "IOT based smart agriculture system." In *2018 International Conference on Wireless Communications, Signal Processing and Networking (WiSPNET)*, pp. 1–4. IEEE, 2018.

Zhao, Wenju, Shengwei Lin, Jiwen Han, Rongtao Xu, and Lu Hou. "Design and implementation of smart irrigation system based on LoRa." *In 2017 IEEE Globecom Workshops (GC Wkshps)*, pp. 1–6. IEEE, 2017.

Chapter 10

Smart irrigation in farming using internet of things

Devesh Kumar Srivastava and Priyanka Nair

MANIPAL UNIVERSITY

10.1 Introduction

Farming in Rajasthan is barely successful. The main cause for the failure of farming techniques is the wrong predictions. These unforeseeable situations are forcing many of the farmers to even take their own lives as they are unable to bear the enormous losses. Hence, we need to keep the essential parameters of agriculture in our mind such as the soil quality and its water content. For a place such as Rajasthan we are already facing the major problem of water scarcity. We have therefore come up with a unique idea of applying internet of things (IoT) in crop monitoring and introducing new techniques of smart farming which could help our farmers to improve their crop production. A solution is offered to the problem of water scarcity so that healthy crops can be cultivated with minimal attention given by the farmer. The main aim of this study is to provide farmers with accurate knowledge of water content in the soil and help these farmers to become financially secure. This work also focusses on designing a cost-efficient way to utilize the in-hand available resources. In Figure 10.1 it is depicted how the proper levels of soil moisture affect crop generation, that is, adequate moisture level in soil helps in producing healthy crops, whereas inappropriate moisture produces unhealthy crops. Our provided solution is responsible for first measuring the soil moisture level and then accordingly releasing water to the fields so that no wastage occurs. It will also alert the farmers about the quality of the soil and the level of water present in the tank. This shows how digital farming has become a boon for the agriculture business. The most vital outcome of this technology is healthy crops with less supervision of fields by the farmers.

In Rajasthan, still the old mundane ways of crop generation are used. We have observed this by constantly monitoring the farms near our university. The uncertainty in crop generation is also because of the lack of knowledge among the farmers; therefore, we need to step up and provide them with the necessary skills. In Rajasthan, as we know and mentioned in the previous references too, water scarcity has always been the major problem that affects soil composition and quality. The lack of providing a feasible solution is still the dire need to keep a check on our farmers' worst situations.

This work aims to bring in notice there has been a big improvement towards the mental health of the farmers who were earlier suffering a lot. It will provide them with a solution to their foremost problem of water scarcity and lay in front of them various methods so as to eradicate the crises completely.

Healthy crops   Unhealthy crops

Figure 10.1 Visualization of crop condition in farming.

The aforementioned method will also help in making these farmers much more financially secure and independent. They will become aware about the proper utilization of the resources in a proper planned manner and also how to develop crops in a cost-effective way.

10.2 Literature survey

IoT technology is growing at a very fast rate and is being used in farming nowadays. The very first conference papers on IoT-based application were published in 2010 (Duan, 2011). Since then, numerous works have been done in IoT field. It is helping in increasing resources. There have been many contributions in this area by some Asian researchers. Agriculture is one of the main research areas which has integrated this technology. Many researches and studies have been done on integrating information technology with agriculture science, water regulation, effect of production on environment, new strategies of production, improvement in soil composition for making it more suitable for arable farming, crop cultivation factors, food supplies preservation and conservation, horticulture with outdoor fruits and vegetables cultivation. With the enhancement of wireless technologies, monitoring and controlling has become easy with new precision agriculture. IoT solutions are surfacing to support the farmers by providing easy solutions to farming challenges.

For the farming solutions, monitoring devices and wireless sensors can be combined under one umbrella that includes predictive analytics on crop production and its growth. Agricultural machinery combined with processing and management of data is showing better outputs in various areas like precision agriculture, livestock farming, arable farming etc. IoT-based technologies are providing great solutions to data-centric agricultural industries. Despite the amazing solutions provided by IoT, many challenges arise when agriculture is combined with these technologies. The further literature survey provides the information of work being done in this field and dwells on the challenges being faced. With the increase in population, the demand for food has been growing tremendously which has led to the demand–supply gap. To fill this demand–supply gap, the production of food needs to increase almost twice the present status (Tripathi *et al.*, 2019). Crop production is also another dominating factor that appends the production of food to the nation's economy. Bioenergy that utilizes fuels from various renewable sources of energy is growing worldwide. Another challenge is that with the increase in population, the use of fossil fuels has increased that has caused a great threat to food security. The smart agriculture industry ensures sustainable use of natural resources and renewable sources of energy due to which it is expected to rise to the market of about 15$ billion in next 5 years. IoT has provided solutions like farming vehicles for reducing the waste and enhancing the production of food. Energy crops are being produced for bioenergy generation.

According to the recent Food and Agriculture Organization reports, it has been predicted that the world may need about 70% more food by 2050 for which there is an urgent need for more farm produce. The solutions provided by IoT are capable enough of filling the demand–supply gap.

With the new technologies, it has become easier to record and measure the environmental changes and find suitable solutions to them. These newly derived wireless networks, devices and sensors consume a lot of energy because of the batteries that are boarded with actuators (Cambra *et al.*, 2017). Autonomous robots are sensor-based and capable of navigation as well as seed sowing along with vision systems, Radio-Frequency Identification (RFID) wireless sensors, GSM (Global System for Mobile Communications), Bluetooth and GPS (Global Positioning System) technology. The data captured from these devices are further transferred for further research. There are many factors that limit the agricultural development such as natural resources availability, agricultural land suitability, increase in population, depletion of biofuels etc. Land employment has significantly declined in the past 10 years. In 1991, land for production of food was approximately 39% of the world's landmass, but later according to the 2013 statistics it reduced to 37% and has been declining continuously with time.

Sensors have been integrated with appliances and water harvesting systems and are helping to track horticulture patterns and livestock breeding. Internet connectivity has made it easier to monitor and control agricultural science growth. Sensing capabilities have facilitated simpler ways to identify the patterns, to predict patterns and hence help in maintaining ecological balance with smart strategies. The IoT works on three layers, first is mainly sensing, second is delivery and third is control (TongKe, 2010). The first layer helps in collecting data from sensors like GPS, RFID etc. Its main job is to collect the data and sense the physical entities and environmental factors. The next layer is an application layer that refers to application delivery in different deployment forms like smart agriculture. The data regarding environmental conditions can be collected and deployed with the application and integrated with the sensors for determining various patterns with different segments such as domestication, horticulture, crop cultivation, irrigation and production of food (Hong, 2011).

With the help of IoT, farmers are now able to navigate, monitor and operate or control mechanisms. In order to improve data management, wireless sensors and aerial drones can gather the data. Existing challenges have been curbed by a solution framework provided by the smart agriculture market. It can further be employed for management of livestock regarding cattle rearing, tracking farm produce, monitoring cattle health and also horticulture development, controlling the production of food and managing crop yield. Augment bioenergy can help to trace food safety. Bluetooth and wireless sensors also help to collect data (Zeinab and Elmustafa, 2017). IoT minimizes human intervention and helps to manage, monitor and control the data. With the help of this technology, the farmers have understood the importance of arable farming and precision agriculture. Smart agriculture and bioenergy have great potential to fill the demand–supply gap that farmers have been facing. Further on the basis of smart agriculture, cost-effective solutions can be suggested. IoT facilitates dynamic tracking and transparency to monitor data to provide significant information in this research area.

10.3 Proposed system

Based on the above concerns we have created a model which is based on IoT for smart farming agriculture activities and follows the below-mentioned objectives:

1. Offer a feasible solution to the farmers of Rajasthan about how to deal with the present water problems.
2. Provide farmers with correct information regarding the soil water content and list them ways for better utilization of water.
3. Produce healthy and improvised quality crops by making farmers aware of the water and moisture content in the soil.
4. Make farmers financially secured by providing them resource utilization methods.
5. Finally, let the digital farming accelerate the agriculture business by constant data monitoring and analyzing. Farmers will be notified with this data for better decision-making.

To achieve the proposed model, three sensors are used:

1. Ultrasonic sensor: This sensor measures the distance with the help of ultrasonic waves. A correct water level that should be present is marked and if the water goes below that point, the sensor notifies the farmer by text or an email to check and refill the tank.
2. Soil moisture sensor: It is used to keep in check the moisture level of soil by sensing. Suppose the moisture is low and soil is dry, then it will automatically supply water to it and when there is enough water present which means the soil is wet then the water pump will stop automatically.
3. Turbidity sensor: The quantitative measure of suspended particles present in a fluid is called turbidity. It is measured as high and low and shown in Figure 10.2.

This sensor helps in measuring the light transmitted in order to determine the turbidity of the water. When there is cloudiness or haziness in the water, the water level rises thus increasing the total suspended solids (TSS). When the specified certain value is crossed in the tank, meaning the tank needs to be cleaned as the water is impure, the system will tend to notify the farmer to ensure the cleanliness of the water tank.

Sending Notification: To inform a farmer about the level of water and its turbidity in the tank, applets are used. Applets are small task applications that perform a specific task and have a well-defined scope. These applets today are often connected with the concept of "If This Then That", commonly referred to as IFTTT.

The term IFTTT is inspired from the use of conditional statements in programming languages that govern logical flow of statements in a program. Primarily, IFTTT is a web-based software service that is responsible for connecting different applications and services run over the internet to trigger automations by making use of applets. For

Figure 10.2 Visualization of turbidity position in water.

instance, in the scope of this project, the farmer will be notified when water reaches low levels or turbidity crosses a threshold value.

Devices used in the system:

1. Arduino UNO

   It is a microcontroller designed to control and sense the metrics associated with electronic device used in the real world. It comes pre-embedded with an ATmega328 microcontroller that provides various timers, regulators, counters and a host of other features.

2. Turbidity sensor

   It is used to ensure water quality and management by focusing a beam of light into the water that is supposed to be monitored.

3. Soil Moisture Sensor

   Soil moisture sensors are used to measure the moisture content present in the soil. These sensors use capacitance at a very high frequency to measure the quantity of water volumetrically against the total quantity of soil present.

(Continued)

4. Ultrasonic sensor

Ultrasonic sensors measure the distance of the targeted object by emitting out sound waves that lie outside the range of frequencies that can be detected by the human ear. Due to its ability of measuring targeted distances, these sensors also find uses in vehicle washing, counting of bottles filled in drink factories and for detecting the presence of people.

5. ESP8266

It is an economical wireless receiver that often finds its use in IoT applications primarily due to the characteristic of providing any microcontroller access to Wi-Fi networks.

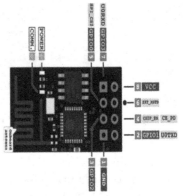

6. Logic level converter

These are bi-directional converters responsible for converting high voltages into low voltages and vice versa. It is important to match the voltage level of other electronic components that are connected on the board of the microcontroller.

7. LED light bulb (used to demonstrate water pump)

8. Jumper wires

9. Resistor

10. Bread board

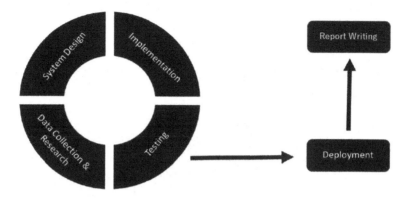

Figure 10.3 Flow diagram.

10.4 Methodology

A six-phase model is proposed to achieve the desired results as described in Figure 10.3.

- Phase 1 consists of gathering of information related to the project. This is essential for defining the project requirements. The right set of tools required to access determine the requirements depending upon the stakeholders involved and the objectives that are to be achieved. Conducting questionnaires and surveys, brainstorming sessions and group discussions are some of the ways to find the key ideas.
- Phase 2 will be designed by keeping in mind the requirements obtained and finalized according to Phase 1.
- In Phase 3, small and independent coding modules are designed which are then combined and integrated into a complete program at the end of this phase or at the beginning of the next. This ensures that the code can be easily debugged and integrated after reviews and revisions. The hardware assembling also takes place in this phase based on the circuit design in previous phase.
- Phase 4 is an evaluation phase. Any problems or issues that might arise due to intermediate steps are checked for and reported. They are fixed by going to the phase from where the issue was originally found.
- Phase 5 marks the completion of product design. The deliverables to be deployed or released are submitted.
- Phase 6 is the final phase where the project is delivered to the client for consumptions. Any issues that may arise are addressed through release of updates and fixes.

Circuit diagram: The system comprises devices like Arduino UNO, turbidity sensor, soil moisture sensor, ultrasonic sensor, ESP8266, LED light bulb, jumper wires, resistor and bread board as shown in Figure 10.4.

10.5 Implementation

```
Void sendNotification()
{
```

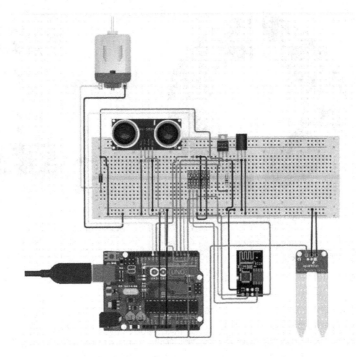

Figure 10.4 Circuit diagram.

```
client.println("GET /trigger/turbidity_low_email");
client.println("Connection: close");
}
```

Four sensors are farmed inside irrigation system.

10.5.1 Automatic watering

The water pumps and soil moisture sensors are the essential encompassment of the framing module. The level of moisture is recorded by the soil moisture sensor. The low-level of moisture in the soil is fulfilled with the sensor's automatic switching of the water pump. The water pump is halted as soon as the sensor senses and records enough moisture in the soil, that is, when the plant receives appropriate amount of water and the soil gets wet.

10.5.2 Water leveling

The water level in the tank is measured by the ultrasonic sensor that is clamped at the top of the water container. The measurement is recorded by measuring the distance to the target. When the water level in the tank goes below a threshold, the farmer is notified via text or an email. On receiving the notification of low water level, the farmer continues refilling the tank to attain the level required.

10.5.3 Water quality

The water's turbidity level, implying cloudiness and haziness attributes, is directly proportional to the amount of TSS. The water tank has turbidity sensors that record the increase in water's turbidity level with increase in TSS. The water is considered to be dirty when the turbidity of water crosses the threshold value and indicates the point. The water which the water cleaning has to be appropriated by the farmers. The farmers receive the notification when the cleaning of the water tank is required.

10.5.4 Notification to the farmer

IFTTT application is employed by the component to send the text notification in the form of SMS to the registered mobile number of the farmer. The notifications, in the form of alerts, update the farmer on the water level and turbidity level of water in the tank.

10.6 Conclusion

10.6.1 Water irrigation based on the crops

Pertaining to different requisites of water by every crop, customized data-based methodology of water irrigation is employed.

10.6.2 Water immersed crops

Adhering to the large water requirement by the crops like rice, the requisite of continued immersion in water still remains a challenge. Sensors capable of remaining immersed in water can be incorporated. Water depth–based technology can also be considered as a solution.

10.6.3 Reliable hardware

Error detection systems are appropriated to identify and prevent the discrepancy and discontinuity in usage. For error-free working, reliable hardware manufacturing is required.

10.7 Future work

The varied requisites of crops in terms of calcium, magnesium, sodium, etc. can be identified and fulfilled. A notification can be sent to the farmers regarding the information about the water contents.

References

Carlos Cambra, Jaime Lloret, and Laura Garcia, "An IoT Service-Oriented System for Agriculture Monitoring," 2017, *IEEE ICC 2017 SAC Symposium Internet of Things Track*. Universidad Politecnica de Valencia, Sandra Sendra, Department of Signal Theory, Telematics and Communications, University de Granada.

Y.E. Duan, "Research on integrated information platform of agricultural supply chain management based on Internet of Things." *Journal of Software* 2011, 6(5), 944–950.

Econ.tu.ac.th/class/archan/Rangsun/EC%20460/EC%20460%20Readings/Global%20Issues/Biofuels/Effect/Bioenergy%20and%20Agriculture.pdf

Li Hong. *IOT and Cloud Computing: Advance Strategic New Industry.* Beijing: Posts & Telecom Press, 2011.

Fan TongKe Modern Education Technology Center of Xi'an International University "Smart Agriculture Based on Cloud Computing and IoT." Shaanxi.

A.D. Tripathi, R. Mishra, K. K. Maurya, R. B. Singh, and D. W. Wilson, "Estimates for world population and global food availability for global health." *The Role of Functional Food Security in Global Health*, (2019, 3–24.

K.A. Zeinab and S.A. Elmustafa (2017). "Internet of Things Applications, Challenges and Related Future Technologies." *World Scientific News* 2017, 67(2), 126–148.

Chapter 11

Automation systems in agriculture via IoT

Shylaja Vinay Kumar Karatangi

G L BAJAJ INSTITUTE OF TECHNOLOGY AND MANAGEMENT

Reshu Agarwal and Manisha Pant

AMITY INSTITUTE OF INFORMATION TECHNOLOGY, AMITY UNIVERSITY

11.1 Introduction

In the modern world, when new technologies are emerging rapidly the need has arisen to be smart in agriculture too. India is a land of agriculture, ranking second worldwide in farm productivity. Before the advancement in the technology, farmers used to manually irrigate their lands which was a time-consuming process as well as required more water supply. Because of consuming more time, it resulted in crop drying. Also, there were many other problems like uneven ploughing problem, intruder attacks and temperature effect on crops. All of these factors result in low productivity, and farmers had to bear lots of losses. Due to the rapid increase in population, the production of food should also be doubled in the next few years. It will be a great challenge to fulfill this need with limited resources.

To increase food productivity, traditional agricultural systems need to be upgraded. There is a dire need to incorporate the upcoming and latest tools and technologies in agriculture which will fulfill the food requirements while making agriculture fully developed and automated. To satisfy the developing need of water systems in India, because of its questionable climatic conditions, it is important to concentrate on supporting water systems and improving the proficiency of the current water system frameworks. The present population of 7.3 billion is expected to reach 9.7 billion in future. The world will consequently need to produce 70% more food in future so that the growing population can be fed. So, farmers need to move to new technologies so that they can meet the growing demand of food production in the world. Making farming smarter is an emerging concept that helps to manage farms by making use of latest technologies to increase the yields along with the quality.

Internet of Things (IoT) is characterized as an arrangement of interrelated processing gadgets, mechanical and advanced machines, items, creatures or individuals that are given a kind of identifiers having capability of data transfer over network with no need of human-to-human or human-to-computer interaction. It can also be thought of as a network that connects things to the internet through particular algorithms provided by remote sensors, microcontrollers and actuators. The exchange of information and communication is done such that intelligent recognition, tracking, positioning, monitoring, management and control can take place. Hence, using smart agriculture the losses in irrigation water and fertilizer can be reduced. It can also provide a favorable climate in terms of humidity and temperature to increase the production of crops in the farm. Real-time monitoring of temperature and humidity plays an important role in various fields of agriculture. Technology related to IoT is designed to guarantee the better growth of crops in optimum conditions. This advancement in technology will lead to progress in agriculture, and it will also be helpful in solving many future problems of agriculture.

Before the evolution of IoT, many researchers worked in the field of precision agriculture and all making use of mainly wireless sensor networks to collect data from various sensors and then return back over the wireless protocol. The data coming from various sensors provide information on different environmental factors. The difficulties faced in analyzing environmental factors could not be provided a satisfactory and correct solution to increase the productivity of crops. There are certain factors which affect productivity if not taken under observation. To resolve these issues, an advanced system should be developed so that productivity and quality can be increased. This system should have features like monitoring, analysis and automation in agriculture. However, these combined and integrated systems are difficult to achieve in agriculture and farming because of many problems. The development of combined and integrated systems has many challenges in logical research, so they are not accessible to farmers. The objective of IoT in smart agriculture is to implement smart system by making use of available resources to enhance productivity. All these features and technologies will provide a support to build a sustainable and productive farming and agriculture system to enter the fourth industrial revolution.

Further, in the coming years combining farming system with IoT can play a key role in increasing the productivity. Production of food needs to be enhanced in future to satisfy the imminent food demand. Water is also a factor of concern as resources are becoming scarcer day by day. Optimal utilization of limited water resources and reducing water wastage is a challenge. In addition, factors that increase yields like an improved ploughing method and reducing crop loss by intruder prevention are also a matter of great concern. At the end integration of IoT technology with automation of farming procedures can solve all the impending agricultural hazards. Using IoT farmers can constantly monitor the field and weather conditions from anywhere making use of internet. So, farmers will be able to observe the important field conditions in real time and then make a comparison with the previous available data on the server. So the focus of applying IoT in agriculture is to make farmers' tasks automated which include tasks like ploughing, irrigation, bird repelling and also soil moisture monitoring, rainwater harvesting, temperature monitoring. The server provides other data that play a very important role for analyzing and monitoring the entire farm on real-time basis. By exposing automation and IoT in agriculture, farmers can save time and money. That will be very helpful for increasing production.

Conventional greenhouses used manual intervention for monitoring and controlling parameters which resulted in misinterpretations and increased costs. However, IoT-driven smart greenhouses can brilliantly screen and control the atmosphere automatically. Different sensors are deployed to quantify the ecological parameters as per the particular necessities of the yield. That information is put away in a cloud-based processing unit for providing information. Importantly, IoT-based farming not only has a huge scope in cultivating activities, it can serve agribusinesses like organic cultivation, family cultivation, including rearing of specific cows, and additionally developing explicit cultures, safeguarding specific or top-quality varieties and so on.

IoT encompasses a system of sensors, actuators, cameras, robots, drones and other associated gadgets which bring a remarkable degree of control to farming. This smart farming revolution reduces pesticide and manure use while increasing their effectiveness. IoT innovations will empower better food traceability, which thus will prompt to increase food safety. Smart farming is mainly based on IoT which integrates smart machines and sensors on farms to make cultivation information driven and data enabled. Smart farming provides a practical approach to increase agricultural production, based on a precise and resource-efficient approach.

11.2 Related research

Muthunpandian *et al.* (2017) proposed an automatic system for crop field monitoring continuously. The model maintains the water levels within the crop field. This developed device is useful in the irrigation system. Another computerized irrigation model was developed by Gutiérrez *et al.* (2014) to optimize water use in agricultural crops. The gadget has a distributed wireless community of soil-moisture and temperature sensors located inside the root sector of the plants. In addition, a gateway unit handles sensor facts, triggers actuators and transmits information to an internet application. Dwarkani *et al.* (2015) proposed a model for smart farming. They used smart sensing system and smart irrigator system with the help of wireless communication technology. In this model, a mechanical bridge slider arrangement is used on which smart irrigator moves. The smart irrigator receives signals with the help of sensors and global system for mobile communication (GSM) module. The crop details are analyzed by a centralized database and transferred to irrigator system to perform automatic actions. Sukumar *et al.* (2018) proposed a model for smart agriculture in which fertilizers are used effectively and it reduces wastage. This system is developed for monitoring crop field using sensors (temperature, soil moisture, humidity and light). The wireless transmission is used to send data received from sensors to a database. This system finds the moisture values from the sensor and turns the lights in the greenhouse ON and OFF based on light sensors and actuators. Mubarak *et al.* (2015) proposed an automated irrigation system which is very economical in terms of power consumption. This system can be implemented in large agricultural fields. With the help of GSM, user can control the motor from anywhere by just sending an SMS. The system is adaptable to manual mode also, if required. This system uses solar energy and it also works in all climate conditions for variety of crops. Fisher and Kebede (2010) proposed a system for monitoring soil moisture, and soil, air and canopy temperatures are measured in cropped fields. Various types of sensors were used in this system for continuous, automated monitoring of crop conditions. Kim *et al.* (2011) developed a system for water management based on wireless sensor networks and a weather station for internet monitoring of drainage water using distributed passive capillary wick-type lysimeters. Mirabella and Brischetto (2011) proposed a model in which farm is made up of several greenhouses. A multiprotocol bridge has been implemented using wireless system so that great flexibility can be provided. Wenshun *et al.* (2013) proposed a monitoring system for the use of fertilizers and pesticides in the field. Many techniques like sensor nodes, mining, wireless networks etc. are used for getting diseases information, so that necessary actions can be taken to save the crop. Wang and Liu (2014) proposed a model for cattle movement in the field using IoT. This system really helps farmers to save their cattle and also their time. Yang *et al.* (2013) proposed a system for CO_2 monitoring on the surface of the soil based on various sensors. They also proposed a system for monitoring humidity, temperature and light intensity. One of the major requirements of smart farming is transmission of quick and reliable information to the farmers. Zhu *et al.* (2014) proposed a system in which data collected by sensors are transmitted to mobile users in a fast, reliable and secure manner. The system capacity is enhanced by means of data encryption and decryption techniques applied to cloud, mobile devices, sensors and cloud gateways. Pratama *et al.* (2019) developed a system for monitoring condition of cattle in real time and to facilitate farmers in terms of monitoring. Various types of sensors for reading body temperature, heart beat rate and movement of cattle are used for collecting data. Based on these data, normal, less normal and abnormal health classification of cattle is done. Nukala *et al.* (2016) discussed the use of IoT technologies in the food supply chain management. This supply chain includes everything from farm to fork like agriculture, food processing, transportation, distribution etc. They discussed various technologies like radio frequency identification, wireless sensor networks, cloud computing and data analytics for maintaining food supply chain so that fresh fruits and vegetables can be delivered. The main factor which limits the productivity of crop yield is drought. Remote sensing is used in rural areas to obtain frequent soil moisture data, which help to analyze the agricultural drought in distant regions. The soil water deficit index is calculated by Martínez *et al.* (2016). Vagen *et al.* (2016) used the moderate-resolution imaging spectro-radiometer sensor to map various soil functional properties to estimate land degradation risk. Santhi *et al.* (2017) developed sensor- and vision-based autonomous robot called Agribot for sowing seeds. The robot is enabled with GPS system to perform on any agricultural land. Further, Karimi *et al.* (2017) proposed non-contact sensing method to determine the seed flow rate. They used sensors with LEDs and signal information linked to the passing seeds to measure the seed flow rate. Cuhac *et al.* (2012) utilized light ward resistors as collector parts for seed stream estimation. They introduced a real-time wireless seed observing framework for seed drill executions. The light transmitting diodes and light ward resistors were introduced on each channel to determine the seed flow with the help of seed counting information. Recently, drones are playing key roles to assist farmers in some areas like soil and field analysis, planting, crop monitoring, irrigation, plant counting and gap detection, spraying the pesticides and detection of plant species etc. (D'Oleire-Oltmanns *et al.*, 2012; Romero-Trigueros *et al.*, 2017; Gnädinger & Schmidhalter, 2017; Faiçal *et al.*, 2017; Hunter *et al.*, 2017).

Farming is important for our survival. It is essential for the growth of a country's economy. It furthermore gives adequate work opportunities to many people. Numerous farmers are still using conventional systems for agricultural practices which result in low yields. Nevertheless, with the advent of computerization customized equipment is left behind, and the yield is improved. So, use of computerized technology is required in the cultivation sector for extending the yield. A large number of papers propose the use of remote sensors for recording data in a brief timeframe to send it to server for processing. The gathered information provides various characteristic elements on which decision-making is done (Anusha *et al.*, 2019).

Monitoring only environmental factors is not adequate for improving the yield of the harvests. There are various components that impact the yield. These components include attack of bugs, etc. There is a likelihood of theft at the time of harvesting. So as to offer responses to each and every such issue, it is imperative to make a facilitated structure which will manage all segments impacting the productivity at each stage like advancement, assembling and post-gathering storing. A system for watching the field is thus required. With IoT, related gadgets have penetrated into our lives, from prosperity and health, home motorization, vehicle and coordination to urban networks. IoT, related devices and automation have found its application in agribusiness. Development has seen different mechanical changes in the latest decades, getting progressively industrialized and advanced. By using distinctive agribusiness inventions, farmers have managed the route toward raising tamed animals and creating yields.

11.3 What is smart agriculture?

There are numerous approaches to allude to current horticulture. For instance, AgriTech refers to the utilization of innovation in agribusiness.

Smart horticulture, then again, is for the most part used to indicate the use of IoT arrangements in farming. The equivalent applies to the smart cultivating definition (Figure 11.1).

Albeit smart farming IoT, just as modern IoT all in all, is not as mainstream as shopper-associated gadgets, yet its market is still extremely powerful. The IoT solutions for horticulture are continually developing. In particular, Business intelligence predicts that the number of farming IoT gadget establishments will hit 75 million by 2020, increasing 20% every year. Simultaneously, the worldwide smart horticulture will significantly increase by 2025, coming to $15.3 billion (contrasted with over $5 billion in 2016). Since the market is as yet evolving, the doors are still open for organizations ready to participate. Building IoT items for farming in the coming years can separate you as an early adopter, and thus assist you in preparing to progress.

11.4 IoT-enabled technologies

IoT has different empowering advances such as wireless sensor networks, cloud computing, big data, embedded systems, security protocols and architectures, protocols

Figure 11.1 Smart agriculture.

empowering correspondence, web administrations, internet and search engines (Anand & Vikram 2016).

Remote sensor network: It consists of different sensors which help in gathering different types of information.

Distributed computing: It provides information to PCs and different gadgets on request.

Big data analytics: In this technology, the data collected from various sources are analyzed to show different patterns, relationships and other important information.

Communication protocols: This is a very powerful technology for connecting different applications. Without this technology IoT-enabled systems cannot work.

Embedded systems: It is a sort of PC system which involves programming to perform different assignments. It contains microchip/microcontroller, RAM/ROM, I/O units and capacity devices.

11.5 Applications of IoT in agriculture

Till now IoT has found applications in numerous industries and agriculture industry is not an exception. Till the end of 2018, the revenue associated with horticulture was USD 1.8 billion internationally and it is still increasing. It is supposed to reach USD 4.3 billion by 2023 with a compound annual growth rate of 19.3%.

Various IoT-enabled machines help in making agribusiness grow fast. It is estimated that the worldwide population is going to hit 9.6 billion by 2050. To take care of this gigantic populace, the agribusiness business is yet limited. Various hurdles in agriculture like weather change and ecological effects can be overcome using IoT. In the late twentieth century, throughout the world the mechanical advancements, for example, tractors and reapers, occurred and were brought into the farming activities. Furthermore, the farming industry depends intensely on inventive thoughts in view of the consistent development.

The industrial IoT has a central purpose of saving costs. In the future, the usage of IoT will increase in the farming tasks. All things considered, few reports tell that the IoT will see a compound yearly improvement with a pace of 20% in agriculture business. Furthermore, the number of related devices (green) will increase from 13 million in 2014 to 225 million by 2024. On account of unavailability of a good communication systems, an IoT provider faces difficulties in remote areas. In such cases, numerous framework providers are making it possible by introducing satellite accessibility and expending cell frameworks (Figure 11.2).

11.6 How did IoT in agriculture make its impression?

Sensors have been introduced to the agribusiness activities a long time ago. The issue with the conventional methodology of using sensor innovation was that it was not ready to handle the live information from the sensors. The sensors used to log the information into their connected memory and later on had the option to process it. With the presentation of industrial IoT in agriculture, undeniably further developed sensors are being used. The sensors are currently associated with the cloud by means of cell/satellite system, which lets us to know the ongoing information from the sensors.

IoT has enabled the ranchers to screen the water tank levels continuously which makes the water system increasingly effective. The progression of IoT innovation in

Figure 11.2 IoT and smart agriculture.

agribusiness tasks has led us to the information of cultivating procedures like how much time and assets a seed takes to turn into a completely developed vegetable. Web of things in agriculture has come up as a second flood of green insurgency. The advantages that the ranchers are getting by adjusting IoT are twofold. It has helped ranchers to diminish their expenses and increment yields simultaneously by providing ranchers with precise information.

11.7 Appropriateness of IoT in agriculture

Smart farming is a powerful arrangement of doing agribusiness and developing nourishment in a practical manner. It is a utilization of actualizing associated gadgets and inventive advancements together into farming. Smart farming significantly relies upon IoT by subsequently reducing the physical work of ranchers and cultivators and in this manner expanding the profitability. With the ongoing agribusiness patterns reliant on farming, IoT has brought enormous advantages like proficient utilization of water and enhancement of information sources.

IoT-based smart farming improves the entire agriculture system by checking the field constantly. With the help of sensors and interconnectivity, the IoT saves the time of the farmers and lessen the excessive usage of water and electricity. It keeps various components like moisture, temperature, soil, etc. under check and gives an impeccably clear continuous information. The advantages of IoT in agriculture are given below.

11.7.1 Atmosphere conditions

Proper atmospheric conditions are essential for cultivation. Besides, having wrong data about atmosphere can create problems for the crop. IoT enables to record and understand the progressing atmospheric conditions as shown in Figure 11.3. Sensors are set inside and outside of the agriculture fields. They

Figure 11.3 IoT in atmosphere conditions.

Figure 11.4 IoT in accuracy farming.

accumulate data from the nature which can be used to pick the right crop which can flourish in the particular climatic conditions. The IoT involves sensors that can distinguish progressing atmospheric conditions like tenacity, precipitation, temperature quite accurately. There are different numbers of sensors to separately record all these parameters and configuration to suit sharp developing necessities. These sensors screen the condition of the harvests and the atmosphere in which it grows. If any unfavorable atmospheric conditions are found, then an alert is sent. The system sets climate conditions so that for long time farmers can get more benefit.

11.7.2 Accuracy farming

Exactness agriculture/precision farming is one of the most praised employments of IoT as shown in Figure 11.4. It makes the cultivation practice more informed and aware with the use of sensors, for instance, trained creature checking, vehicle following, field observation and stock watching. The goal of precision farming is to analyze the data, gathered by methods for sensors, to react accordingly. Precision farming makes farmers collect data with the help of sensors and inspect that information to take sharp and fast decisions. There are different precision developing methods like water framework, tamed creature administrators, vehicle following and much more for each different activity. In precision farming, one can obtain soil conditions and other related parameters to extend the operational efficiency. Not simply this, it can moreover distinguish the progressing working conditions of the related activities to perceive water and supplement level.

11.7.3 Smart greenhouse

To make our farms clever, IoT has enabled atmosphere stations to normally modify the environmental conditions according to a particular crop. With the introduction of IoT in greenhouses, human intervention has become limited according to the need of the entire system. For example, solar-powered fueled IoT sensors are fabricated present-day for various applications. These sensors assemble and transmit the progressing data which helps in checking the nursery state precisely constantly. With the help of the sensors, the water usage and nursery state can be checked using SMS alerts. Modified and clever water frameworks are developed with the help of IoT. These sensors help provide information about weight, dampness and temperature and light levels as shown in Figure 11.5.

11.7.4 Data analysis

The traditional database framework needs more stockpiling for the information gathered from the IoT sensors as shown in Figure 11.6. Cloud-based information stockpiling and a start-to-finish IoT platform assumes a significant job in the agricultural business framework. These frameworks are evaluated to assume a significant job with the end goal that better exercises can be performed. In the IoT world, sensors are essential for gathering information for an enormous scope. The information is broken down and changed to important data utilizing investigation apparatus. The data can

Figure 11.5 IoT in greenhouse.

Figure 11.6 IoT data analysis.

Figure 11.7 IoT used in drones.

be analyzed to provide useful information for better yields. After doing data analysis, the patterns found help in forecasting climate conditions for harvesting.

11.7.5 Information analytics

Technological progressions have nearly changed the horticultural tasks, and the presentation of farming automatons is the slanting interruption. The ground and aerial automatons are utilized for appraisal of harvest wellbeing, crop observing, planting, crop splashing and field investigation (Figure 11.7).

11.8 Benefits of brilliant cultivating: how's IoT forming farming?

Innovations and IoT can possibly change farming in numerous perspectives. To be specific, there are five different ways in which IoT can improve farming:

Figure 11.8 Crop harvesting.

- Data and huge amounts of information are gathered by savvy agribusiness sensors, for example, climatic conditions, soil quality, yield's development progress or cows' wellbeing. This information can be utilized to follow the condition of the business such as staff execution, hardware proficiency and so on.
- Better authority over the internal procedures helps reduce unforeseen dangers. The capacity to anticipate the yield creation permits to get ready for better item dispersion. If we know precisely how much yields, as shown in Figure 11.8, are going to be reaped, it can be ensured that the harvest will not remain unsold.
- Cost and waste decrease because of good monitoring and estimation. Having the option to perceive any irregularities in the harvest development, will allow the option to relieve the dangers of losing the yield.
- Business proficiency increases through procedure robotization. By utilizing smart gadgets, it can mechanize different procedures, for example, water system, treating or bug control.
- Good quality and volumes of crops are harvested. With automation there are better expectations of yield as well as quality.

Therefore, these components can in the long run lead to higher income. Since it has been laid out how IoT can be profitably applied in agribusiness, it should be investigated how the recorded advantages can discover their application, all things considered.

11.9 Things to consider before building up smart cultivating arrangement

Obviously, the utilization cases for IoT in farming are endless. There are numerous ways in which sensors and smart gadgets can increase the harvest and income.

Nonetheless, farming IoT applications improvement is not so simple. There are difficulties that should be considered before putting IoT into smart cultivation.

11.9.1 Equipment

To assemble IoT equipment for horticulture, sensors need to be purchased (or custom ones made). The decision will rely upon the sorts of data it needs to gather. Regardless, the nature of sensors is critical to the accomplishment of the problem: it will rely upon the precision of the gathered information and its unwavering quality.

11.9.2 Brain

Information investigation ought to be at the center of each savvy farming arrangement. The gathered information itself will be of little assistance if it is not comprehensible. In this way, we need to have ground-breaking information investigation abilities and apply prescient calculations and AI so as to get noteworthy bits of knowledge dependent on the gathered information.

11.9.3 Maintenance

Support of the equipment is a test that is of essential significance for IoT items in agribusiness, as the sensors are regularly utilized in the field and can be effortlessly harmed. Therefore, it should be ensured the equipment is solid and simple to maintain. Otherwise sensors would need supplanting more regularly than required (Figure 11.9).

11.9.4 Mobility

Smart cultivating applications should be hand crafted for use in the field. A business should have the choice to provide the information on the spot or remotely by methods

Figure 11.9 Maintenance.

for a phone or work station. Additionally, each related device should have enough remote range to talk with various devices and send data to the central server.

11.9.5 Structure

To guarantee that smart cultivating application performs well (and to ensure it can deal with the information load), it needs a strong internal framework. Moreover, internal frameworks must be secure. Neglecting to appropriately ensure security of the framework just increases the likeliness of somebody breaking into it, taking the information.

11.10 Web of food/farm 2020

In the event that we have IoT in farming and the internet of medical things (IoMT) in medicine field, why not have one for nourishment? The European Commission venture "Internet of Food and Farm 2020" (IoF2020), a piece of Horizon 2020 Industrial Leadership, investigates through research and standard meetings the capability of IoT advancements for the European nourishment and cultivating industry.

IoT has encouraged the conviction that a savvy system of sensors, actuators, cameras, robots, rambles and other associated gadgets will bring a remarkable degree of control and computerized dynamic to agribusiness, making conceivable a suffering biological system of development.

11.11 Third green revolution

Brilliant farming and IoT-driven horticulture are making ready for what can be known as a third green revolution. Following the plant reproducing and hereditary qualities, the third green revolution is assuming control over horticulture. That insurgency draws upon the joined use of information-driven examination advancements, for example, exactness cultivating gear, IoT, "large information" investigation, unmanned aerial vehicles (UAVs or automatons), mechanical technology and so on.

Later on, pesticide and manure use will drop while the effectiveness will rise. IoT innovations will empower better nourishment, which in turn will prompt expanded sanitation. It will likewise be valuable for the earth, as it progressively increases effective utilization of water, or improvement of medicines and information sources.

In this manner, brilliant cultivating has a genuine potential to convey a progressively profitable and maintainable type of horticultural creation, in view of an increasingly exact and asset proficient methodology. New ranches will at long last understand the unceasing dream of humanity. It will take care of our populace, which may explode to 9.6 billion by 2050.

Table 11.1 shows the development of IoT-based selection in agriculture part from forecast of years 2000–2016 and 2035–2050.

11.12 Conclusion and future scope

IoT-empowered farming has helped actualize present-day mechanical answers for tried and true information. This has helped conquer any hindrance in growing

Table 11.1 IoT and agriculture current scenario and future forecasts

Year	Data analysis
2000	525 million farms associated with IoT
2016	540 million farms associated with IoT
2035	780 million farms would be associated with IoT
2050	2 billion farms are probably going to be associated with IoT

quality and good amount of yield. Information acquired by different sensors for continuous use or capacity in a database guarantees quick activity and less harm to the harvests. With consistent start to finish clever activities and improved business process execution, produce gets prepared quicker and arrives at stores in quickest time conceivable. So, today it has been figured out how farming fields are profited by IoT frameworks. Thus, IoT will become familiar in the forthcoming instructional exercises. Moreover, stay tuned to learn all the more intriguing things that you can do with this innovation. Future work would be centered more around expanding sensors to get more information particularly for pest control and by additionally incorporating GPS module to improve this agriculture IoT technology to undeniable agricultural precision.

References

Anand, N. & Vikram, P. (2016). Smart Farming: IOT based smart sensors agriculture stick for live temperature and moisture monitoring using Arduino, cloud computing and solar technology, International Conference on Communication and Computing Systems. DOI: 10.1201/9781315364094-121.

Anusha, A., Guptha, A., Rao, G. S., & Tenali, R. K. (2019). A model for smart agriculture using IOT, *International Journal of Innovative Technology and Exploring Engineering*, 8(6), 1656–1659.

Cuhac, C., Virrankoski, R., Häninen, P., Elmusrati, M., Hööpakka, H., & Palomäki, H. (2012). Seed flow monitoring in wireless sensor networks, *2nd Workshop on Wireless Sensor Systems*, USA, pp. 70–73.

D'Oleire-Oltmanns, S., Marzolff, I., Peter, K.D., & Ries, J.B. (2012). Unmanned Aerial Vehicle (UAV) for Monitoring Soil Erosion in Morocco, *Remote Sensing*, 4(11), 3390–3416.

Dwarkani, C. M., Ram, G. R., Jagannathan, S., & Priyatharshini, R. (2015). Smart agriculture system using sensors for agricultural task automation, *Proceedings of IEEE International Conference on Technological Innovations in ICT for Agriculture and Rural Development*, Chennai, Tamilnadu, pp. 49–53.

Faiçal, B. S., Freitas, H., Gomes, P. H., Mano, L. Y., Pessin, G., Carvalho, A. C. P. L. F, Krishnamachari, B., & Ueyama, J. (2017). An adaptive approach for UAV-based pesticide spraying in dynamic environments, *Computers and Electronics in Agriculture*, 138, 210–223.

Fisher, D. K. & Kebede, H. A. (2010). A low-cost microcontroller-based system to monitor crop temperature and water status, *Computers and Electronics in Agriculture*, 74(1), 168–173.

Gnädinger, F., & Schmidhalter, U. (2017). Digital counts of maize plants by unmanned aerial vehicles (UAVs), *Remote Sensing*, 9(6), 1–15.

Gutiérrez, J., Medina, J. F. V., Garibay, A. N., & Gándara, M. A. P. (2014). Automated irrigation system using a wireless sensor network and GPRS module, *IEEE Transactions on Instrumentation and Measurement*, 63(1), 166–176.

Hunter, M. C., Smith, R. G., Schipanski, M. E., Atwood, L. W., & Mortensen, D. A. (2017). Agriculture in 2050: recalibrating targets for sustainable intensification, *BioScience*, 67(4), 386–391.

Karimi, H., Navid, H., Besharati, B., Behfar, H., & Eskandari, I. (2017). A practical approach to comparative design of non-contact sensing techniques for seed flow rate detection, *Computers and Electronics in Agriculture*, 142, Part A, 165–172.

Kim, Y., Jabro, J. D., & Evans, R. G. (2011). Wireless lysimeters for realtime online soil water monitoring, *Irrigation Science*, 29(5), 423–430.

Martínez, F. J., González-Zamora, A., Sánchez, N., Gumuzzio, A., & Herrero-Jimenez, C. M. (2016). Satellite soil moisture for agricultural drought monitoring: assessment of the SMOS derived Soil Water Deficit Index, *Remote Sensing of Environment*, 177, 277–286.

Mirabella, O. & Brischetto, M. (2011). A hybrid wired/wireless networking infrastructure for greenhouse management, *IEEE Transactions on Instrumentation and Measurement*, 60(2), 398–407.

Mubarak, S., Khan, S., Sahana, N., & Sujatha, S. (2015). Automated irrigation system using wireless sensor networks and GSM module, *International Journal of Advance Research in Science and Engineering*, 4(1), 188–198.

Muthunpandian, S., Vigneshwaran, S., Ranjitsabarinath, R. C., & Reddy, M. K. (2017). IOT based crop-field monitoring and irrigation automation, *International Journal of Advanced Research Trends in Engineering and Technology*, 4(19), 450–456.

Nukala, R., Panduru, K., Shields, A., Riordan, D., Doody, P. & Walsh, J. (2016). Internet of Things: a review from 'Farm to Fork', *Proceedings of Irish Signals and Systems Conference, Londonderry*, UK, pp.1–6.

Pratama, Y. P., Basuki, D. K., Sukaridhoto, S., Yusuf, A. A., Yulianus, H., Faruq, F., & Putra, F. B. (2019). Designing of a smart collar for dairy cow behavior monitoring with application monitoring in microservices and internet of things-based systems, *International Electronics Symposium*, Indonesia, pp. 527–533.

Romero-Trigueros, C., Nortes, P.A., Alarcón, J. J., Hunink, J. E., Parra, M., Contreras, S., Droogers, P., & Nicolas, E. (2017). Effects of saline reclaimed waters and deficit irrigation on Citrus physiology assessed by UAV remote sensing, *Agricultural Water Management*, 183, 60–69.

Santhi, P. V., Kapileswar, N., Chenchela, V. K. R., & Prasad, C. H. V. S. (2017). Sensor and vision based autonomous AGRIBOT for sowing seeds, *Proceedings of International Conference on Energy, Communication, Data Analytics and Soft Computing*, Chennai, pp. 242–245.

Sukumar, P., Akshaya, S., Chandraleka, G., Chandrika, D., & Kumar, R. D. (2018). IOT based agriculture crop-field monitoring system and irrigation automation, *International Journal of Intellectual Advancements and Research in Engineering Computations*, 6(1), 377–382.

Vagen, T.-G., Winowiecki, L. A., Tondoh, J. E., Desta, L. T., & Gumbricht, T. (2016). Mapping of soil properties and land degradation risk in Africa using MODIS reflectance, *Geoderma*, 263, 216–225.

Wang, X. & Liu, N. (2014). The application of Internet of things in agricultural means of production supply chain management, *Journal of Chemical and Pharmaceutical Research*, 6(7), 2304–2310.

Wenshun, C., Lizhe, Y., Shou, C., & Jiancheng, S. (2013). Design and implementation of sunlight greenhouse service platform based on IOT and cloud computing, *Proceeding of the IEEE International Conference on Measurement, Information and Control*, China, pp. 141–144.

Yang, H., Qin, Y., Feng, G., & Ci, H. (2013). Online monitoring of geological CO_2 storage and leakage based on wireless sensor networks, *IEEE Sensors Journal*, 13(2), 556–562.

Zhu, C., Wang, H., Liu, X., Shun, L., Yang, L. T., & Leung, V. C. M. (2014). A novel sensory data processing framework to integrate sensor networks with mobile cloud, *IEEE Systems Journal*, 10(3) 1125–1136.

Chapter 12

A complete automated solution for farm field and garden nurturing using Internet of Things

S. Siva Kumar, A. M. Senthil Kumar, T. Rajesh Kumar, N. Sree Ram, V. Krishna Reddy and P. Sivakumar

KONERU LAKSHMAIAH EDUCATION FOUNDATION

12.1 Introduction

In developing countries like India, irrigation of crops and plants with the required amount of water is the fundamental problem. The main economy of the country depends on agriculture. Due to isotropic climatic conditions of the country, different regions need varied amount of water. Most of the lands in the country are dry and due to lack of rain a lot of lands are changing slowly into the unirrigated part. Unplanned use of water in irrigation may lead to a significant waste of water. The farmer uses manual control to irrigate the land at regular intervals in this era. Because of these reasons, the land gets more water in some parts while water reaches other parts late. In order to overcome this problem, we are using an automatic drip irrigation system.

This system supplies water to the root zone of the plant and the crops are irrigated only when there is an intense requirement of water. The different parts of the country have variation in soil, and different amount of nutrients are required by the crops and plants. To overcome this situation, we have employed the pH level and electrochemical sensor (Saranu *et al.* 2018) in the system. The entry of unauthorized animals and birds must be monitored and controlled to overcome the damaging of food crops. This can be done by the involvement of the Infra Red (IR) sensor and with a random buzzer in the system. The system also supports inspection of the crops at regular intervals through a camera. It is required to overcome the effect of various insects that influence the growth of the crop. The proposed system sends information about the sensed information to the user mobile phone through Internet of Things (IoT) network. The entire irrigation field and garden status are monitored, controlled and implemented through this system.

The IoT is used to interconnect everyday objects, tools, devices or computers together into one system. The network may vary in size, place and time. Every object is tagged by a sensor. The sensor data are stored in the cloud. The cloud acts as a repository of sensor data. These data are retrieved from the cloud and processed and sent to the user in the form of image or text. The IoT gateway is nothing but Raspberry Pi, which is used to connect the sensor network and the cloud. The gateway gathers data from sensors, processes the request from remote users, uploads the data to the cloud and instructs the user based on the situations. The involvement of IoT in the fields and gardens gives better management, improved plant disease prevention ability, increase in production by the integrated use of soil moisture sensor, temperature sensor, pH level sensor, water level sensor, electrochemical sensor and IR sensor.

The article is divided into many sections. Section 12.2 gives the literature reviews of various implementations of IoT in the agricultural field. The proposed architecture is given in Section 12.3. It includes the discussion of the architectural diagram, the workflow of the architecture and a sequence diagram. Section 12.4 analyzes the result of the system. Finally, Section 12.5 ends with a conclusion.

12.2 Literature survey

Al-Ali *et al.* (2015) presented a wireless irrigation system for a home garden that can be integrated with the home control systems. The system consists of a master station and a set of slave nodes connected with a wireless microcontroller. Each slave node contains a soil moisture sensor, a temperature sensor, a water valve, a ZigBee transceiver and a microcontroller. The slave microcontroller reads the temperature and soil moisture from the trees and grass, then constructs a frame for transmission. The frame is forwarded via ZigBee ad hoc network to the master station.

Al-Bahadly and Thompson (2015) gave a system that measures the soil moisture and determines whether soil requires water or not. The system simulates a garden sprinkler that utilizes a dual output tap timer which consists of two water valves. A Teensy 2.0 microcontroller reads the two moisture-sensing circuits and controls the water valves. The system is more successful because of the more reliable use of the soil probes. Tripathy *et al.* (2015) gave a system which uses soil moisture, light and temperature sensors to decide (Nimmagadda *et al.* 2020) the amount of water to be supplied to the plants. The microcontroller is configured by embedded C and Python programming languages. The sensor data are displayed to the user using graphical user interface. The user will control and monitor the system remotely.

Suzuki *et al.* (2013) used a support vector machine to classify the sensor data that are received from the agricultural system. A cloud system is used to support and store the sensor values. Even if users do not know the irrigation, this system is expert in irrigating agricultural field properly. Nandurkar *et al.* (2014) gave irrigation using IoT that saves electricity and water and thereby reduces the labor cost. The initial cost of the system was very low so that it can be used by the small farmers. This system also increases the yield of the crops.

Zhao and Ye (2008) transmitted the sensor data to the user in the form of short message using Global System for Mobile (GSM). GSM/General Packet Radio Service (GPRS) is a simple and more convenient way of communication to the end system than the internet. This network is available almost all the time even in the case of failure of the internet. The information can be sent or received with high privacy in the network. Abbas *et al.* (2014) has given an irrigation process which is controlled by a wireless network for watering the agricultural fields using valves. This implementation is more efficient in terms of time and quantity of water that is supplied to the fields.

Shinde and Gatlawar (2015) developed a wireless sensor network in an agro-based automation system that helps to analyze and compare the data using fuzzy logic. The monitored analog parameters are transmitted to the other side where they can be read and controlled by a set of points. Mekala and Viswanathan (2017) discuss the remote maintenance of the agriculture field. It monitors the temperature, humidity and water level in the soil. The remote-control method also includes animal or bird scaring, spraying and weeding. Based on the real-time sensor data it makes an intelligent decision-making for smart (Kasthuri *et al.* 2017) irrigation.

Li *et al.* (2014) describes the multiple access methods such as GSM, Wi-Fi, GPRS, Third Generation (3G), Enhanced Data GSM Environment (EDGE), Local Area Network (LAN) and so on. The data are stored locally in this system. The IoT gateway uses a microcontroller unit as STM32 and embedded operating system as μC/OS-III. The gateway is reliable, extendible and compatible as demonstrated by the application. The gateway is used to realize the real-time detection and control of the greenhouse and improved the power of the intelligent and automated system (Soniya *et al.* 2017) for greenhouse monitoring.

12.3 Proposed architecture

Figure 12.1 shows the proposed architecture of the complete irrigation and garden nurturing system. Raspberry Pi is the main controller that collects data from all sensors to the cloud database. The soil moisture sensor is used to sense the water content in the soil, which is placed in the soil. The level sensor is placed nearer to the root of the crops which computes the range of the water level, and if the level of water is low then the motor is switched on. pH sensor is also placed, which is to measure the salt content in the water. It is used to reduce fertilizer usage.

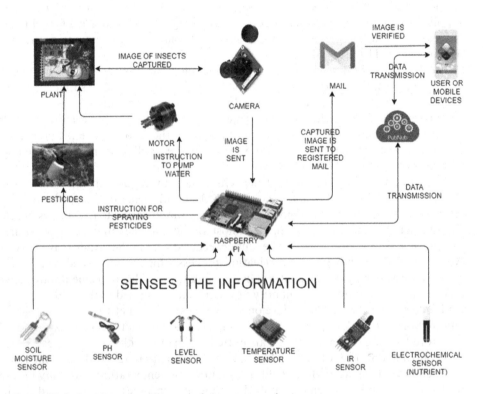

Figure 12.1 Architecture diagram of the proposed system.

The temperature sensor (Jennifer *et al.* 2017) is used to check the weather conditions. IR sensor is included to monitor the entry of cows, goats, birds, ducks, etc. The electrochemical sensor is used to monitor the nutrient level in the soil the plant requires. If the nutrient content is below the expectation, then fertilizers are provided. The complete data are transmitted which the sensor senses. Then the data are collected by Raspberry Pi and transmitted to the mobile phone. The plants and crops are monitored for infection through any foreign organism, the image is captured, mailed or sent to the farmers or gardeners on their mobile phones.

12.3.1 Work flow

Figure 12.2 shows the sequence diagram of the proposed system. In the proposed system IoT plays a major role. The purpose of the IoT in this system is to share the data with the users. Raspberry Pi (Kaveeya *et al.* 2017b) is connected with a Wi-Fi module. The farmers or gardeners have to register their mobile with Raspberry Pi for communication and control. The information of the soil is transmitted to the Wi-Fi network through the signal conditioning circuit of various sensors. The physical information of the soil such as soil moisture, the temperature is sent to the Wi-Fi, it is stored in the cloud and shared with the user using IoT.

The moisture sensor checks the water content in the soil. If it is less than the reference value, the information is passed on to the farmer. The farmer turns on the motor

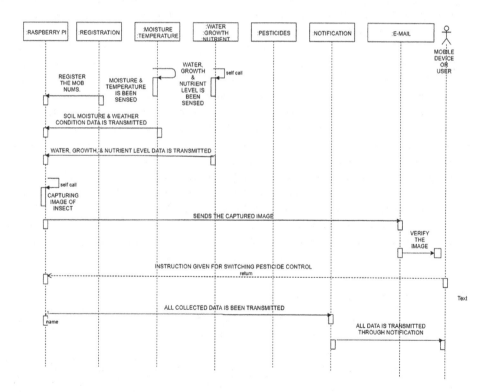

Figure 12.2 Sequence diagram of the proposed system.

to irrigate the crops through the mobile phone (Brindha *et al.* 2017; Neeraj Kumar *et al.* 2018a; Maurya and Kumar *et al.* 2017; Kumar *et al.* 2011; Kumar *et al.* 2018). The information about height of the crops and chlorophyll in the plant is also shared to the user's personal computer with an internet connection or smartphone. If the moisture content of the soil is lower than the reference value, then the command from the user device is transmitted to the field section through IoT, then the irrigation system is activated and the water is supplied to the field. If the moisture content of the soil reaches the span value, then the irrigation system is deactivated, this information is also transmitted to the user (Neeraj Kumar *et al.* 2018b, 2018c, 2018d). This is a chain process of this particular proposed irrigation system.

12.3.2 Intrusion detection sub-system

In this sub-system, IR sensor, camera and buzzer play a major role. IR sensor is activated to capture moving object in the fields. Once an object is detected, the camera will be activated by the Pi. The camera captures the moving object into the fields. The moving objects may be a human being, cow, goat, buffalo, pigeons, dove, crow, etc. The captured image is sent to the PC for image classification. The classification algorithm of four-layer convolutional neural network (CNN) (Sivakumar S. et al. 2019) detects whether the image is human or non-human. If it is a non-human, the buzzer will be initiated by Pi. It will give an alert to the animals to move away from the fields. The system will update the status of the entry to the mobile phone (Neeraj Kumar *et al.* 2010; Raj *et al.* 2016) of the gardener or farmer. Figure 12.3 shows the working principle of intrusion detection sub-system.

12.3.3 Water supply sub-system

In this sub-system, moisture sensor, temperature sensor, level sensor and motor play a major role. Soil moisture sensor (Sagar *et al.* 2017) checks the moisture content in the soil. If soil moisture is low, it continuously checks the moisture in the soil. Otherwise, the temperature sensor is activated. Pi checks if the temperature is less than 35°C. If the temperature is less than 35°C, the threshold level will be set to normal. Otherwise, the threshold level will be set to double the value of normal. Pi switches on the motor. The level sensor is also activated by the Pi to check the threshold level. If the threshold level is reached, Pi switches off the motor. Pi sends the status information to the farmer/gardener's mobile phone. Figure 12.4 shows the working principle of water supply sub-system. The temperature and humidity output of a day are shown in Figure 12.5.

12.3.4 Nutrition sub-system

This sub-system consists of motor, pH sensor and rainfall sensor. The rainfall sensor is activated by the Pi to sense the fall of rain. If rain is falling, the motor (Saranu *et al.* 2018) is running or not is checked by the Pi. If the motor is running, Pi will turn off the motor and it activates the pH sensor to sense the nutrition content in the soil. If the pH value is less than 7, acidic nature message is sent to the mobile. If the pH value is greater than 7, the alkaline message is sent. Otherwise, the neutral message is sent

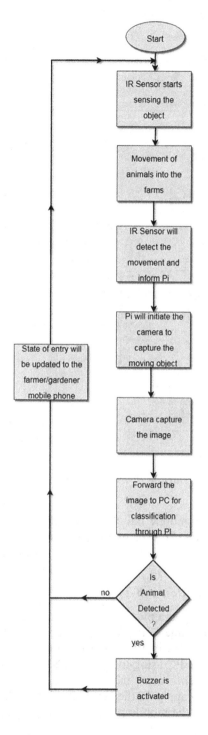

Figure 12.3 Working principle of intrusion detection sub-system.

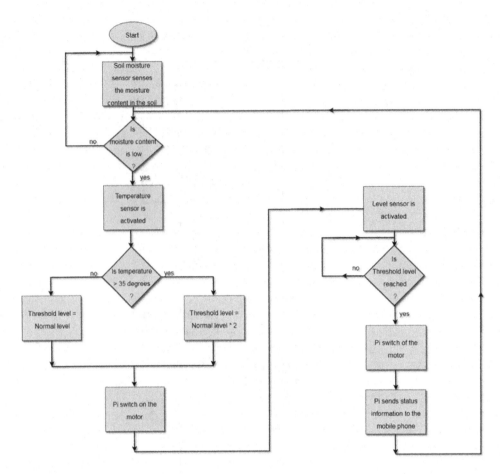

Figure 12.4 Working principle of water supply sub-system.

to the farmer/gardener's mobile phone. Figure 12.6 shows the working principle of nutrition sub-system. The pH level and motor runtime output of the nutrition system is shown in Figure 12.7.

12.3.5 Infection detection sub-system

In this sub-system, the camera plays the main role. Pi activates the camera (Kaveeya *et al.* 2017a) at a specific time of the day. The camera captures the leaf images of the plant. It passes the captured image to the PC for classification. Whether the leaf is infected is checked in the PC using four-layer CNN. If it is infected, the leaf image is passed on to the gardener/farmer's mobile phone. The gardener views the image and decides the pesticides to apply. If the leaf is not infected, the status of the infection alone will be passed on the mobile phone. Figure 12.8 shows the working principle of infection detection sub-system. The output of infection and intrusion detection is shown in Figure 12.9.

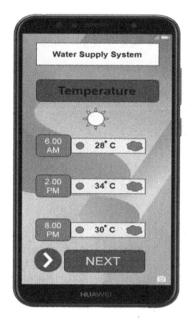

Figure 12.5 Temperature and humidity of the water supply system.

12.4 Experiment and result

12.4.1 Proposed architecture

Figure 12.10 shows the proposed architecture of the four-layer deep CNN. It is used to classify the entry of non-humans in the intrusion detection system (P Arokianathan et al. 2017) and the presence of the disease in the leaves of the paddy crops. The architecture includes four layers of CNN followed by batch normalization with a ReLu activation function. The normalization is included to stabilize, to increase the speed and finally to increase the performance of the network. Each CNN layer includes a kernel size of 3, 3. A dropout of 0.25 is included in the second, third and fourth CNN layer to reduce the over fitting of the features (Sivakumar and Rajalakshmi 2017) across the network. The first and second CNN layers contain a filter size of 32, while the third and fourth layers include filters of sizes 64 and 128, respectively. A max pooling layer is included in the second and fourth layers of CNN. This pooling layer is used to reduce the dimensionality of the input feature and thereby increase the computation speed. The four layers of CNN act as a feature extraction phase of the network. The second phase of the network includes three dense layers used for the classification of the images based on the feature extracted. The three dense layers consist, respectively, of 512, 128 and 4 as their output units. The first and second dense layers consists of ReLu activation function, batch normalization and a dropout of 0.5. The last dense layer consists of softmax activation function, loss as categorical cross-entropy and optimizer as Adam. The softmax function is used as output in the neural network to get "k"

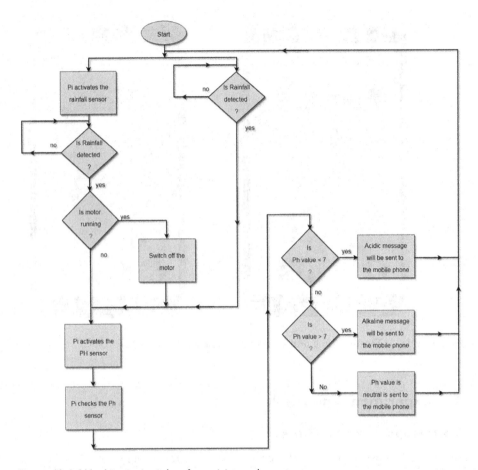

Figure 12.6 Working principle of nutrition sub-system.

different possible outcomes. The categorical cross-entropy is a loss function applicable to an instance that belongs to one class only. Adam is a combination of Stochastic Gradient Descent (SGD) with momentum and RMSprop.

12.5 Experimental setup and dataset

For experimental setup, we used the system of configuration: Intel Core i3, 2.40 GHz, 6 GB RAM with 70 GB hard disk in Ubuntu 16.04 environment. All the experiments were performed using Python 3.5 version, Tensorflow 1.3 version, Keras 2.0.8 version and cuDNN. Keras is a deep learning library which contains all deep learning models (Kumar T. *et al.* 2020). Pickle 1.3 library is used to save and load the preprocessed object. The sklearn library is used for retrieving the confusion matrix. The convolution layer is created by the function Conv2D(). The normalization, pooling layer and dropout are created by the function BatchNormalization(), MaxPooling2D() and

Figure 12.7 pH level and motor runtime of nutrition system.

Dropout(), respectively, defined in the Keras library. The classification phase is constructed by a function dense() from the Keras library.

The confusion_matrix() function from sklearn library is used to obtain the confusion matrix. From this matrix, we can get true positive (TP), true negative (TN), false positive (FP) and false negative (FN) values. The metrics like accuracy, precision, recall and F1 score can be obtained from the confusion matrix values. Accuracy is the only metrics used in the experiment to measure the performance of the classification. It is shown in Equation (12.1). The dataset is taken from UCI (Rice-Disease-DataSet – GitHub) and GitHub (Rice Leaf Diseases DataSet – UCI). Bacterial leaf blight (BLB), brown spot (BS), blast (BL) and leaf smut (LS) are the four diseases of paddy taken from these two sources for training and testing as shown in Figure 12.11. The first source gives BLB, BS and LS of 40 images each. The second source gives BLB, BL and BS of 96, 80 and 100 images, respectively. From the two sources, the total images obtained for the experiment are BLB, BS, BL and LS of 136, 140, 80 and 40 images, respectively. The dataset is divided into 80% for training and 20% for testing.

$$\text{Accuracy} = \frac{(TP + TN)}{(TP + TN + FP + FN)} \tag{12.1}$$

12.6 Result and discussion

The quality of the image captured using the camera is affected by the flash, frequency distortion, ambient light and intensity of the camera. The above factors are considered

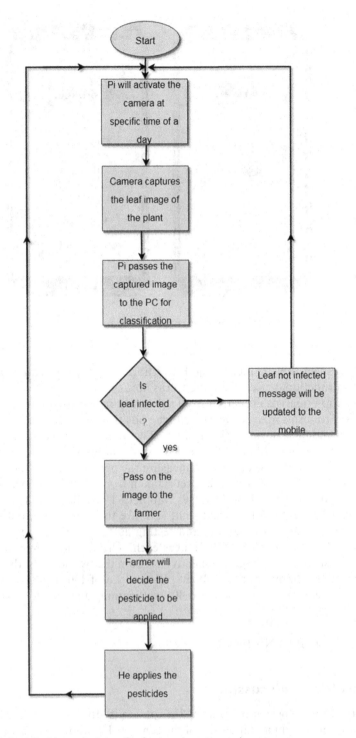

Figure 12.8   Working principle of infection detection sub-system.

Figure 12.9 Output of infection detection and intrusion detection system.

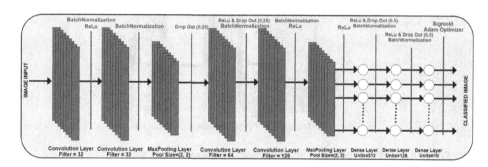

Figure 12.10 Proposed architecture diagram of four-layer deep CNN.

as noise. It can be removed from the image using a deep CNN (Andrey *et al.* 2017). An enhanced image for the brown spot disease of paddy leaves is shown in Figure 12.12. Image augmentation method artificially creates multiple images for training by combining the different processes such as rotation, translation, shearing, flips, etc. Augmentation is done to increase the number of images in the smaller-size dataset to improve the performance of training. The ImageDataGenerator() is a function from Keras library used to generate augmented images artificially.

The experiment is conducted for different epochs like 1, 10, 20, 30 and 40. Table 12.1 shows the training and validation accuracy for different epochs by a four-layer deep CNN on the paddy disease dataset. Figure 12.13 shows the comparison (Apoorva Sindoori *et al.* 2017) of training accuracy with validation accuracy. In the figure, 1, 2, 3, 4 and 5 on the x-axis correspond to 1, 10, 20, 30 and 40 epochs, respectively.

(a) Brown Spot   (b) Bacterial Blight

(c) Rice Blast   (d) Leaf Smut

Figure 12.11 Four different diseases of the paddy crop. (a) Brown spot. (b) Bacterial blight. (c) Rice blast. (d) Leaf smut.

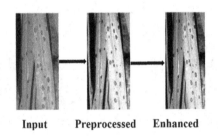

Input   Preprocessed   Enhanced

Figure 12.12 Preprocessing of paddy leaf for brown spot disease.

Table 12.1 Training and validation accuracy for different epochs

Epochs	Training accuracy	Validation accuracy
1	90.42	86.78
10	92.845	91.267
20	93.761	92.071
30	95.367	93.891
40	98.893	97.847

The four-layer CNN gives a high training and validation accuracy of 98.893% and 97.847%, respectively. Table 12.2 shows the validation accuracy of non-augmented and augmented images for different epochs. In Figure 12.14, 1, 2, 3, 4 and 5 on x-axis correspond to 1, 10, 20, 30 and 40 epochs, respectively. In Table 12.2, the augmented dataset gives a high validation accuracy of 98.43% at 40 epochs than non-augmented dataset.

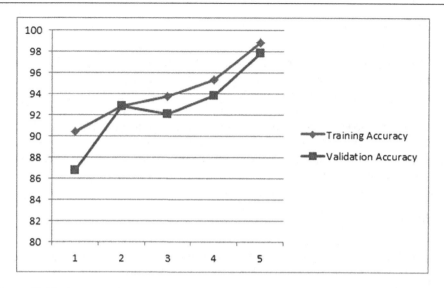

Figure 12.13 Comparison of training accuracy with validation accuracy.

Table 12.2 Non-augmented and augmented accuracy for different epochs

Epochs	Non-augmented	Augmented
1	86.78	87.89
10	91.267	92.46
20	92.071	93.87
30	93.891	95.35
40	97.847	98.43

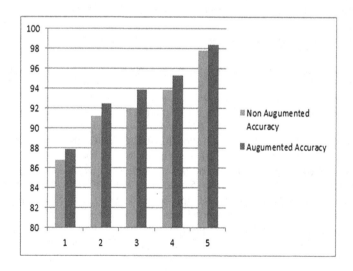

Figure 12.14 Comparison of non-augmented and augmented validation accuracy.

12.7 Conclusion

Improvement in production of crops is a major challenge in a developing country like India. Novel smart technologies should be taken under agriculture stream to lead to green population country. So, we proposed a complete automated technology based on IoT for the farm fields and gardens. The plants and crops are monitored and controlled without effort from humans. This system has integrated the four subsystems to provide a unified approach for gardening/framing of plants and crops. This system has the ability for further improvement by incorporation of new self-learning techniques. It will be easy to incorporate it into IoT to understand the behavior of sensor-collected data and take individual or autonomous decisions.

References

A. H. Abbas, M. M. Mohammed, G. M. Ahmed, E. A. Ahmed & R. A. A. A. Abul Seoud, (2014), Smart watering system for gardens using wireless sensor networks, *2014 International Conference on Engineering and Technology (ICET)*, Cairo, pp. 1–5, doi: 10.1109/ICEngTechnol.2014.7016780.

A. R. Al-Ali, M. Qasaimeh, M. Al-Mardini, S. Radder & I. A. Zualkernan, (2015), ZigBee-based irrigation system for home gardens, *2015 International Conference on Communications, Signal Processing, and their Applications (ICCSPA'15)*, Sharjah, pp. 1–5, doi: 10.1109/ICCSPA.2015.7081305.

I. Al-Bahadly & J. Thompson, (2015), Garden watering system based on moisture sensing, *2015 9th International Conference on Sensing Technology (ICST)*, Auckland, pp. 263–268, doi: 10.1109/ICSensT.2015.7438404.

Ignatov Andrey, Kobyshev Nikolay, Timofte Radu, Vanhoey Kenneth & Van Gool Luc, (2017), DSLR-quality photos on mobile devices with deep convolutional networks, *Proceedings of the IEEE International Conference on Computer Vision*, Venice, pp. 3297–3305, doi: 10.1109/ICCV.2017.355.

K. B. Apoorva Sindoori, L. Karthikeyan, S. Sivakumar, G. Abirami & Ramesh Babu Durai, (2017), Multiservice product comparison system with improved reliability in big data broadcasting, *IEEE 2017 Third International Conference on Science Technology Engineering & Management (ICONSTEM)*, doi: 10.1109/ICONSTEM.2017.8261256.

P. Arokianathan, V. Dinesh, B. Elamaran, M. Veluchamy & S. Sivakumar, (2017), Automated toll booth and theft detection system, *IEEE 2017 Technological Innovations in ICT for Agriculture and Rural Development (TIAR)*, pp. 84–88, doi: 10.1109/TIAR.2017.8273691.

S. Brindha, D. Deepalakshmi, T. Dhivya, U. Arul, S. Sivakumar & K. N. Kannan, (2017), ISCAP: intelligent and smart cryptosystem in android phone, *2017 International Conference on Power and Embedded Drive Control (ICPEDC)*, pp. 453–458, doi: 10.119/ICPEDC.2017.8081132.

J. Jennifer, M. N. Marrison, J. Seetha, S. Sivakumar & P. Saravanan (2017), DMMRA: dynamic medical machine for remote areas, *2017 International Conference on Power and Embedded Drive Control (ICPEDC)*, pp. 467–471, doi: 10.1109/ICPEDC.2017.8081135.

R. Kasthuri, B. Nivetha, S. Shabana, M. Veluchamy & S. Sivakumar, (2017), Smart device for visually impaired people, *Third International Conference on Science Technology Engineering & Management (ICONSTEM)*, pp. 54–59, doi: 10.1109/ICONSTEM.2017.8261257.

Gokula S. Kaveeya, S. Gomathi, K. Kavipriya, A. K. Selvi & S. Sivakumar, (2017a), Automated unified system for LPG using load sensor, *2017 International Conference on Power and Embedded Drive Control (ICPEDC)*, pp. 656–660, doi: 10.1109/ICPEDC.2017.8081133.

Neeraj Kumar, Abhishek K. Pandey & R. C. Tripathi, (2010), A framework to prevent mobile sinks accessing by unauthorized nodes in WSN, *Special issue on MANET, IJCA (USA)*, pp. 13–17.

Neeraj Kumar, Amit Kumar & Deepak Chaudhary, (2011), A novel approach to use nano sensor in WSN applications, *International Journal IJCA*, (USA), Volume 14, No. 2, pp. 31–34.

N. Kumar, A. Agrawal & R. A. Khan, (2018), Smartphone with Solar Charging Mechanism to Issue Alert During Rainfall Disaster, In: Panda B., Sharma S., Roy N. (eds) *Data Science and Analytics. REDSET 2017. Communications in Computer and Information Science*, vol 799. Springer, Singapore.

Neeraj Kumar, Alka Agrawal & R. A. Khan, (2018a), Cost estimation of cellularly deployed IoT enabled network for flood detection, *Iran Journal of Computer Science*, Volume 2, No. 1, pp. 53–64.

Neeraj Kumar, Alka Agrawal & R.A. Khan, (2018b), Parameters to design an expert system to reduce risk for rainfall induced disaster: flood perspective, *International Journal of Pure and Applied Mathematics (IJPAM)*, Volume 120, No. 6, pp. 1051–1065.

Neeraj Kumar, Geeta Arora, Alka Agrawal & R. A. Khan, (2018c), TSD algorithm to design CA based expert system for pipelining to stop urban flood, *International Journal of Engineering and Technology (UAE)*, SPC, Volume 7, No. 3.1, pp. 56–62.

Neeraj Kumar, Alka Agrawal & R. A. Khan, (2018d), A novel drainage system using cellular automata to avoid urban flood, *International Journal of Applied Evolutionary Computation (IJAEC)*, IGI Global, Volume 9, No. 2, pp. 38–51.

T. Rajesh Kumar, L. S. Videla, S. SivaKumar, A. G. Gupta & D. Haritha, et al., "Murmured Speech Recognition Using Hidden Markov Model," 2020 7th International Conference on Smart Structures and Systems (ICSSS), Chennai, India, 2020, pp. 1–5, doi: 10.1109/ICSSS49 621.2020.9202163.

Guohong Li, Wenjing Zhang & Yi Zhang, (2014), A design of the IOT gateway for agricultural greenhouse, *Sensors & Transducers*, Volume 172, No. 6, pp. 75–80.

Awadhesh Kumar Maurya & Neeraj Kumar, (2017), Localization problem in disaster management smartphone application, *International Journal of Advanced Research in Computer Science*, Volume 8, No. 9, pp. 177–180.

M. S. Mekala & P. Viswanathan, (2017), A novel technology for smart agriculture based on IoT with cloud computing, *2017 International Conference on I-SMAC (IoT in Social, Mobile, Analytics and Cloud) (I-SMAC)*, Palladam, pp. 75–82, doi: 10.1109/I-SMAC.2017.8058280.

S. R. Nandurkar, V. R. Thool & R. C. Thool, (2014), Design and development of precision agriculture system using wireless sensor network, *2014 First International Conference on Automation, Control, Energy and Systems (ACES)*, Hooghy, pp. 1–6, doi: 10.1109/ACES.2014.6808017.

S. Nimmagadda, S. Sivakumar, N. Kumar & D. Haritha, (2020), "Predicting Airline Crash due to Birds Strike Using Machine Learning," 2020 7th International Conference on Smart Structures and Systems (ICSSS), Chennai, India, 2020, pp. 1–4, doi: 10.1109/ICSSS49621.2020.9202137.

R. T. Raj, S. Sanjay & S. Sivakumar, (2016), "Digital licence mv", *2016 International Conference on Wireless Communications", Signal Processing and Networking (WiSPNET)*, pp. 1277–1280, doi: 10.1109/WiSPNET.2016.7566342.

Rice-Disease-DataSet – GitHub https://github.com/aldrin233/RiceDiseases-DataSet

Rice Leaf Diseases Data Set – UCI Machine Repository https://archive.ics.uci.edu/ml/datasets/Rice+Leaf+Diseases

S. V. Sagar, G. R. Kumar, L. X. T. Xavier, S. Sivakumar & R. B. Durai, (2017), SISFAT: Smart irrigation system with flood avoidance technique, *2017 Third International Conference on Science Technology Engineering and Management (ICONSTEM)*, pp. 28–33, DOI 10.1109/ICONSTEM.2017.8261252.

Prithvi Nath Saranu, G. Abirami, S. Sivakumar, Kumar M. Ramesh, U. Arul & J. Seetha, (2018), Theft detection system using PIR sensor, *IEEE International Conference on Electrical Energy Systems (ICEES)*, pp. 656–660, doi: 10.1109/ICEES.2018.8443215.

Reena P. Shinde & Yogesh N. Gatlawar, (2015), Automated environment monitoring and control system for agro-based industries using wireless sensor networks, *International Journal of Research in Advent Technology* Special Issue National Conference "ACGT 2015", pp. 13–14.

S. Sivakumar & R. Rajalakshmi, (2017), Comparative Evaluation of various feature weighting methods on movie reviews, *Springer 2017 International Conference on Computational Intelligence in Data Mining (ICCIDM)*, pp. 721–730, doi: 10.1007/978–981-10–8055-5_64.

S. Sivakumar, R. Rajalakshmi, K. B. Prakash, B. R. Kanna & C. Karthikeyan, (2019), Virtual vision architecture for VIP in ubiquitous computing. In: Paiva S. (eds) *Technological Trends in Improved Mobility of the Visually Impaired. EAI/Springer Innovations in Communication and Computing.* Springer, Cham, pp. 145–179, doi: 10.1007/978-3-030–16450-8_7.

V. Soniya, R. Swetha Sri, K. Swetha Titty, R. Ramakrishnan & S. Sivakumar, (2017) Attendance Automation Using Face Recognition Biometric Authentication, *IEEE 2017 International Conference on Power And Embedded Drive Control (ICPEDC)*, pp. 122–127, doi: 10.1109/ICPEDC.2017.8081072.

Y. Suzuki, H. Ibayashi & H. Mineno, (2013), An SVM based irrigation control system for home gardening, *2013 IEEE 2nd Global Conference on Consumer Electronics (GCCE)*, Tokyo, 2013, pp. 365–366, doi: 10.1109/GCCE.2013.6664857.

A. K. Tripathy, A. Vichare, R. R. Pereira, V. D. Pereira & J. A. Rodrigues, (2015), Open source hardware based automated gardening system using low-cost soil moisture sensor, *2015 International Conference on Technologies for Sustainable Development (ICTSD)*, Mumbai, pp. 1–6, doi: 10.1109/ICTSD.2015.7095915.

Y. Zhao & Z. Ye, (2008), A low cost GSM/GPRS based wireless home security system, *IEEE Transactions on Consumer Electronics*, Volume 54, No. 2, pp. 567–572, doi: 10.1109/TCE.2008.4560131.

Chapter 13

Machine intelligence techniques for agricultural production

Case study with tomato leaf disease detection

Mihir Narayan Mohanty

ITER, SIKSHA 'O' ANUSANDHAN

13.1 Introduction

Agriculture has been the sole reason for the transformation of nomadic life to the settled life of humans. With passing times, it not only became the source of the main food supply for the ever-growing population but also acted as raw material feeder industry for most of the industries, leading to the overall development of humankind. While the industries grew and accommodated new technologies, the farmer himself had no or limited access to data on climate, water supply, energy availability, demands in the market, movements of the pests like locusts or even bigger animals and plant diseases. This all has led to the fact that even if farming as an industry is the source of livelihood it has lagged miles behind in terms of integration with data analytics and other relevant techniques, affecting the overall yield.

Smart agriculture encompasses the use of many on-field sensors like cameras, soil moisture sensors, temperature sensors, pH sensors, gas and smoke sensors and light intensity sensors. The data from the sensors can be collected either through a direct connection to the computer or can be saved to the cloud. If the data are saved to the cloud, it may be available as a reference as well. The data related to the market can be collected from various sources available on the internet.

The data collected from various sources can now be put through rigorous analysis. This is where machine learning comes into actual play. The data can be standardized and finally put through various learning algorithms for clustering, classification, prediction and finally control and monitoring of many variables. The control here implies controlling various relays, motors and other actuators for manipulating the immediate environment. This is aimed to help the farmer in proper maintenance of the crops and the livestock with a lesser amount of personal intervention. The data and various interpretations thus generated can be used to help other farmers as well in real time. Technological advancements have proved themselves to be highly reliable in increasing the yield. If technological advancements are aided with accurate analysis of data, then agriculture like any other sector can be organized and made profitable.

This chapter explains the related state-of-the-art technologies to date. Then a case study is proposed for detecting various diseases in tomato leaf. Several machine learning algorithms were considered for this purpose. Deep learning algorithm is found to be one of the effective techniques though different approaches have been used. Also, neural network plays a vital role to perform the task.

13.2 Related literature

Different agriculture frameworks dependent on internet of things (IoT) are largely encouraged for food creation and decrease the utilization of assets like water. Bu and Wang (2019) have introduced a brilliant farming IoT framework dependent on deep reinforcement learning which incorporates four layers, specifically rural information assortment layer, edge processing layer, rural information transmission layer and distributed computing layer. Their framework uses data methods, particularly manmade consciousness and distributed computing, with horticultural creation to expand food production. Exceptionally, the most developed manmade consciousness model, profound fortification learning, was incorporated in the cloud layer to settle on quick keen choices, for example, deciding the measure of water to be flooded for improving the harvest. A few support learning models with their expansive applications were additionally introduced. The shortage of fresh water assets in the world has created a requirement of their ideal usage. IoT arrangements, in light of the application of explicit sensors' information securing and smart preparing, are crossing over the gaps between the digital and physical universes. IoT-based smart water system frameworks can help in accomplishing ideal water-asset use in the accuracy cultivating scene. In Goap, Sharma, Shukla, and Krishna (2018), an open-source innovation-based smart framework has been proposed to anticipate the water supply necessities of a field employing different soil factors like soil drizzle, temperature and natural surroundings flanking the climate gauge information from the internet. The total framework has been created and conveyed on a pilot scale, where the sensor hub information was remotely gathered over the cloud utilizing web administrations and an online data perception and choice-supportive network gives the constant data experiences dependent on the investigation of sensors information and climate estimate information. The framework has an arrangement for a shut circle control of the water to understand a completely self-governing water system. The framework was completely practical and satisfactory. Most useful part of IoT is to join each part of the world in such a way that people can control them by means of internet. Besides, these items give standard reports of their present status to its end client. Despite the fact that IoT ideas were projected a few decades back, it might be wrong to cite that this stretch has turned into a target for building up correspondence between factors. Researchers (Khanna and Kaur 2019) have assessed the commitments made by different scientists and academicians in the course of recent years. Besides, existing difficulties confronted while performing horticultural exercises have been featured alongside upcoming research work for preparing new ideas in this area to survey the flow of IoT and to additionally enhance them with all the more motivating and creative thoughts.

Different artificial intelligence (AI) strategies that supervise the problems are introduced by IoT by taking into consideration smart city areas as the primary use case in Mahdavinejad *et al.* (2018). The key commitment of their work was the prologue of a technical categorization of AI calculations illuminating how different approaches were applied to the information so as to extract significant amount of data. The difficulties of AI in favor of IoT information investigation were examined. A support vector machine (SVM) model process traffic information was introduced for a progressive point-by-point exploration. Wireless sensor networks (WSNs) are utilized for the genuine execution of the IoT in smart horticulture, savvy structures, smart urban communities and online modern observing applications. For the most part, conventional WSN hubs are controlled by restricted vitality limit, non-battery-powered batteries. The WSN lifetime relies on working cycle, sort of utilization arrangement and battery charge level. Researchers (Sharma, Haque, and Jaffery 2019) have proposed an imaginative answer for the constrained vitality accessibility structure issue by using the solar-powered battery charging of WSN hubs. Notwithstanding, there are numerous difficulties in solar-based power gathering like discontinuity of accessible force, sun-oriented vitality expectation, warm issues, sun-oriented board transformation proficiency and other natural issues. The target of their work was to boost the WSN lifetime utilizing sunlight-based power gathering procedure. Agriculture segment is advancing with the emergence of the data and correspondence innovation. Endeavors are being made to improve the efficiency and diminish misfortunes by utilizing the cutting-edge innovation and hardware. As the majority of the ranchers are ignorant of the innovation and most recent practices, numerous master frameworks have been created to encourage the ranchers. Be that as it may, these master frameworks depend on the put away information base. A specialist framework dependent on the IoT that can exploit the information assembled constantly was proposed in Shahzadi, Tausif, Ferzund, and Suryani (2016). It will assist in taking proactive and defensive actions to bound the disasters due to ailments and creepy crawlies/bugs. Farming is at the core of all professions for creating nations, and having creating advances, the purpose ought to be practical and effectual. The projected arrangement in Katyal and Pandian (2020) incorporates temperature sensors for streamlining water use and yield, and radar sensors for observing any intrusion in the ranch. The arrangement was intended to provide information to shrewd farming incorporating water irrigation system with predictable checking for climate conditions in the present and future. A intrusion-checking framework was introduced which can locate creatures or bugs attacking the fields. Their arrangement uses solar-powered batteries as backup power, so a sun-oriented board is utilized in the smaller than expected ranch. The primary target was to give a similar investigation of savvy ranches to ordinary homesteads which utilize AI calculations progressively to handle issues of water and vitality. The IoT and AI have been progressing mechanical purposes in every single manner, and discovering its way in horticulture is as yet troublesome because of the costs which probably would not be reasonable for a rancher. Customary farming frameworks require gigantic measures for field watering. A smart water system framework that assists ranchers with watering their horticultural fields utilizing global system for mobile communication (GSM) was proposed in Krishnan *et al.* (2020). Their framework gives affirmation messages about the activity statuses, for example, mugginess level of soil, temperature of general condition and status of engine with respect to power. Fluffy rationale controller was utilized to register input parameters (e.g., soil dampness, temperature and moisture) and to deliver yields of engine status. Their framework additionally turns off the engine when there is rainfall. The correlation was made between the proposed framework, dribble water system and manual flooding. The examination results demonstrate that water and electricity preservation was achieved through the proposed shrewd water system framework. Agribusiness has given a significant wellspring of nourishment for people over a huge number of years, including the improvement of proper cultivating techniques for various sorts of yields. The rise of IoT advances can possibly screen the rural condition to guarantee top-notch solutions. There is an absence of innovative work according to smart and sustainable farming, joined by complex hindrances emerging from the fracture of rural procedures, for example, the control and activity of IoT/AI machines, information sharing and the execution, interoperability and a lot of information examination and capacity. Alreshidi (2019) investigated existing IoT/AI advancements received for smart sustainable agriculture (SSA). He recognized IoT/AI specialized design fit for supporting the improvement of SSA stages. Just as adding to the flow of information, their examination surveys innovative work in SSA and gives an IoT/AI design to build up an SSA stage. As of late, different deep learning techniques have been generally contemplated and applied in different fields including horticulture. Scientists in the fields of farming regularly use programming structures without adequately looking at the thoughts and instruments of a method. It was proposed in Zhu *et al.* (2018) that a compact synopsis of significant deep learning calculations, including ideas, constraints, usage, preparing procedures and model codes, help scientists in farming. Research on deep learning applications in horticulture was summed up and examined, and future open doors were talked about, which was required to help specialists in farming. Likewise other asset compelled settings past IoT by creating fundamentally better list items when contrasted with Bing's L3 ranker when the model size is limited to 300 bytes was proposed. So as to meet the prerequisites of insight development in present day farming, proposition to build up a shrewd rural information cloud stockroom under the Hadoop modest PC bunches and the IoT, utilizing progressed astute nursery innovation to gather data through different kinds of detecting gadgets, to understand the opportune, dependable and productive capacity, access just as examination of the information was structured in Jiang, Wang, and Qi (2019). Distributed storage, information the executives, thinking and understanding component, man-made brainpower and different advances are applied to build up a total arrangement of new yield reproducing process, in this way shaping a coordinated model of new assortments rearing innovation to address the issues of farming examination applications. With the advent of IoT and industrialization, the advancement of information technology (IT) has prompted different examinations in industry as well as in horticulture. Particularly, IoT innovation can overcome the limitations of wired correspondence frameworks and can expect horticultural to undergo IT improvement with computerization of agrarian information assortment. In Yoon, Huh, Kang, Park, and Lee (2018), shrewd homestead framework utilizing low-force Bluetooth and low-power wide area networks communication units was developed. Moreover, the framework is used for checking and controlling farms utilizing the message queuing telemetry transport (MQTT) strategy, which was an IoT devoted convention, in this way upgrading the chance of improvement of agrarian IoT. Shrewd agribusiness or savvy cultivating involves the use of IoT for harvesting crops with the capability of reducing work and assets, controlling watering and treatment, gathering accurate data about planting conditions. Changmai, Gertphol, and Chulak (2018) created a smart hydroponic ranch utilizing IoT innovation to explore its advantages. Lettuce was picked as the testing crop. Savvy homesteads can screen the developing condition of plants and modify supplement arrangement, air temperature and air mugginess as indicated by the sensors. The fundamental goal of the smart horticultural framework is to improve the yield of the field. In Varman, Baskaran, Aravindh, and Prabhu (2017), two standards are mentioned: (i) anticipating the reasonable harvest for the following yield; (ii) controlling the water system of the field. The aforementioned objective is accomplished by intermittently checking the fields. The observing procedure includes gathering data about the dirt parameters of the field. A remote sensor arrangement (WSN) was built to gather these information and have data of the past stored to the cloud. This transferred information frames the reason for examination. Through experimentation, long short term memory (LSTM) systems were seen as the appropriate solution. The deduced outcomes were contrasted and the ideal qualities and the most appropriate harvest was informed to the client through SMS. Faltering is a major issue in dairy business, ranchers are not yet ready to satisfactorily accept it on account of the high introductory expenses and complex hardware, and thus, an IoT application that uses AI and information examination methods was proposed in Byabazaire, Olariu, Taneja, and Davy (2019). The portability information from the sensors, appended to the front leg of each dairy animal, was collected at the mist hub to shape time arrangement of social exercises. The information was processed in the cloud and inconsistencies were sent to rancher's cell phone utilizing message pop-ups. The application and model naturally gauge and can assemble information consistently to such an extent that dairy animals can be observed every day. This implies there is no requirement for crowding the bovines, moreover the grouping strategy utilized proposes another methodology of having an alternate model for subsets of creatures with comparable movement levels rather than a one-size-fits-all methodology. It likewise guarantees that the custom models are powerfully modified as climate and ranch conditions change. The underlying outcomes show that we can foresee problems 3 days before it tends to be outwardly caught by the rancher with a general exactness of 87%. This implies the creature can either be disengaged or treated quickly to stay away from any further impacts of weakness. IoT assumes a major significant job in rural industry as of late so as to offer a help to ranchers, for example, checking temperature, dampness and water gracefully, and furthermore early illness observing and identification framework. To give a smart cultivating arrangement, an IoT framework with a bot notice on tomato developing stages was proposed in Kitpo, Kugai, Inoue, Yokemura, and Satomura (2019). The tomato dataset was acquired from Shinchi AgriGreen, the tomato nursery in Fukushima, Japan. They have prepared and tried the profound learning model to recognize the organic product proposition district. At that point, the recognized areas were arranged into six phases of organic product development utilizing the obvious frequency as an element in SVM grouping with the weight exactness of 91.5%. Savvy farming is a promising IoT application area in the Industry 4.0 system. Atmospheric changes influence the common assets turning them into a genuine issue in the rural and food creation setting. Nonstop observing of ecological parameters and rural procedures robotization can prompt assets improvement. Both the reasonable model and the structure of Solarfertigation, an IoT framework, are explicitly intended for brilliant farming as proposed in Valecce, Strazzella, Radesca, and Grieco (2019). Specifically, the imagined arrangement had the option to identify probably the most important territory parameters to take care of a dynamic procedure that drives mechanized preparation and water system subsystems. Moreover, to accomplish self-maintainability, Solarfertigation was fueled by a photovoltaic plant. The key highlights of Solarfertigation are delineated all through that commitment together with its starter model execution. Salam and Shah (2019) introduced an IoT innovation research and development guide for the field of horticulture (PA). Numerous ongoing functional patterns and the difficulties have been featured. Some significant targets for coordinated innovation research and training in agribusiness are depicted. Compelling IoT-based communication and detecting ways to deal with difficulties in agribusiness were introduced. Common logical progressions have engaged most recent innovative approaches to show up. An easy and adaptable solution for controlling and screening farming devices, fundamentally a DC-driven machine for water system using plug gadgets, was structured in Das, Deb, Biswal, and Das (2019). The advanced plugging system was a force switch which can be switched by means of Wi-Fi or any other communication convention. The proposed solution is straightforward, requires minimal effort, simple to move and simple to control. The first idea utilized here was consistent control of DC motor with H-bridge circuit utilizing power as in insulated gate bipolar transistor (IGBT). Here, essentially the DC appliance was selected in light of the fact that there were just bunch techniques for parametric control and that the proposed strategy was dependable and can even work with the littlest advance conceivable. Geographical information system (GIS) was utilized for receiving the limited satellite information of the farmland, whereas with the assistance of on-field sensors and nearby information processing units, real-time information can be accessed. Loads of devices and procedures are accessible in the farming segment. IoT assumes the significant job of improving profitability, proficiency and worldwide market. It additionally decreases human intercession, cost and time which are central point in farming. IoT can be characterized as a framework which interrelate gadgets, objects, machines (like mechanical and advanced) and living creatures. In this way, so as to expand efficiency, IoT works with horticulture to get brilliant cultivation. How IoT has revolutionized savvy cultivation was talked about (Bhagat, Kumar, & Kumar, 2019). The present agribusiness industry is information focused, exact and more intelligent than any time in recent memory. The capability of remote sensors and IoT in agribusiness is featured in Ayaz, Ammad-Uddin, Sharif, Mansour, and Aggoune (2019). IoT gadgets and communication methods used in farming applications are discussed in detail. How IoT is helping the cultivators through all the yield stages, from planting until collecting, pressing and transportation, is also discussed. Moreover, the utilization of unmanned vehicles for crop observation and other ideal applications, for example, streamlining crop yield, was thought of. Best in class IoT-based designs and stages utilized in horticulture were additionally featured. Lastly, in light of that careful survey, they have distinguished momentum and future patterns of IoT in farming. Farming area possesses 25.9% of the world business. The interest for food creation is quickly expanding with the expansion of total populace. Building up the current agrarian framework by joining present-day advances will assist with coordinating this expanding request. A framework to ideally control the atmosphere and water system in a nursery by checking temperature, soil dampness, moistness and pH through a cloud-associated portable robot which can distinguish the undesirable plants utilizing picture handling was proposed in Dharmasena, de Silva, Abhayasingha, and Abeygunawardhana (2019). A controller that can control the temperature, water system and humidifiers in the nursery depending on the sensor readings was introduced. The versatile robot explores through a predefined guide of the nursery and gathers soil tests to perform estimations while locally available sensors gather the surrounding atmosphere information. A camera mounted on the robot will catch the plant and identify weeds dependent on the shading and the surface of the leaves. Changes in precipitation and atmosphere have become extremely irregular in the most recent decade. Indian ranchers need to utilize counterfeit strategies called savvy horticulture to handle these. Researchers in Mukherji, Sinha, Basak, and Kar (2019) have proposed a savvy farming framework utilizing IoT with remote systems administration idea, MQTT, to screen the continuous agrarian condition. Quickly creating IoT was applied in numerous remote situations. A remote monitoring station (RMS), which consolidates web and remote correspondences, was recommended. The significant point was to gather quick information of agrarian field air utilizing MQTT, CC3200 by Texas Instruments and Sensors, and send it to the RMS with the goal that the ranchers will be educated about appropriate upkeep of the fields and subsequently will keep up perfect yield. Farming is the foundation of each nation. It creates all the essential needs, for example, wheat, rice, natural products, grains which are devoured by a human for regular endurance. Thus, it is significant for the nation to create and continue a profitable rural framework. As request is expanding for food, food security and incrementing the yield at a higher rate is essential by simultaneously safeguarding the environment. Along these lines, the innovations in the horticultural area might be fused to upgrade food supplies and creation. Sensor innovation utilized in this area is profoundly successful, accurate and beneficial for horticulture (Ramdinthara & Bala, 2019). To rebuild Japan's declining food independence rate and revive the field of horticulture, the idea of smart farming and urban agribusiness are being executed. In Veloo, Kojima, Takata, Nakamura, and Nakajo (2019), a framework for acquiring composite development information in different situations and harvests focused for home nurseries and paddy fields was proposed. A detecting framework comprising of IoT-based advances was structured and acknowledged to guarantee the consistent development of harvests in ideal conditions. With this, progress will be made in deciding the effective development conditions for AI and in discovering answers for future issues of farming. Major to this IoT transformation is the appropriation of minimal effort, long-run correspondence advances that can undoubtedly manage an enormous number of associated detecting gadgets without expending over the top force. In Citoni, Fioranelli, Imran, and Abbasi (2019), a survey and examination of long-range wide area network–empowered IoT application for smart horticulture was introduced. Long-range wide area network restrictions and bottlenecks were discussed with specific spotlight on their effect on agrarian applications. Deep learning is a promising methodology for smart horticulture, as it maintains a strategic distance from the work concentrated element building and division-based limit. In Ale, Sheta, Li, Wang, and Zhang (2019), authors have first proposed a densely connected convolutional network–based exchange learning technique to distinguish healthy plants from diseased plants, which hopes to run nervous servers with expanded processing assets. To lessen the size and calculation cost of the model, they have additionally used DNN to demonstrate and diminish the size of information. The proposed models were prepared with various-sized pictures to estimate the suitable size of the information pictures. Analysis results were given to assess the proposed models dependent on genuine world dataset, which show the proposed models can precisely distinguish plant illness utilizing low computational assets.

These days, savvy agribusiness using remote correspondence is supplanting the wired framework which was hard to introduce and look after. Another plan for IoT application, which uses various advances to introduce another model for pragmatic usage in the IoT, was presented in Li *et al.* (2019). That plan can settle another technique to take care of issues in market demand, precision in operation and oversight. Moreover, proposed configuration can be utilized more and help ranchers, croppers and individuals to build up their business. AI has generally been exclusively performed on servers and superior machines. Along these lines, with the present progression of these gadgets; as far as the preparing power, vitality stockpiling and memory limit is taken into consideration, the open door has emerged to remove incredible incentive in having on-gadget AI for Internet of Things (IoT) gadgets. Actualizing AI-enabled gadgets have colossal potential which is still in its beginning times. In Yazici, Basurra, and Gaber (2018), stage forward has been considered to comprehend the achievability of running AI calculations. In that particular work, an implanted variant of the Android working framework is intended for IoT gadget improvement using both preparation and deduction on a Raspberry Pi platform. Three unique calculations, random forests, SVM and multi-layer perceptron, individually have been tried utilizing ten various ous informational indexes on the Raspberry Pi to profile their performance regarding speed (preparing and surmising), precision and device utilization. According to the tests, the SVM calculation ended up being somewhat quicker in deduction and productive in power utilization, yet the random forest calculation displayed the most noteworthy accuracy. Advances in the IoT are assisting in making water utilization more economical in horticulture industry. Another topology of sensor hub dependent on the utilization of economical and profoundly productive segments, for example, water level, soil dampness, temperature, stickiness and downpour sensors, was proposed in Khoa, Man, Nguyen, Nguyen, and Nam (2019). Moreover, to ensure great execution of the framework, the pre-owned transmission module was dependent on long-range wide area network innovation. The structure of the circuit was advanced by consolidating two layers and executing programming. The general sensor arrangement was created and tried in their examination lab. Trial results were delivered by testing sensors and their communication adequacy, and were in this manner approved in the field through a 1-week estimation battle. Throughout the years, Machine learning procedures have been actualized to improve information preparing pace and result in IoT gadgets. Probably the most widely recognized machine learning calculations incorporate Bayesian statistics, neural networks, K-nearest neighbors (KNN), SVM, K implied clustering, genetic algorithms, choice trees, principal component analysis, random forest and regression analysis. The previously mentioned calculations have been utilized for grouping for different uses, for example, discovering flaws in the information, improve speeds by framing bunches of comparative information focuses. Machine learning can be utilized in different fields, for example, to foresee heart ailments, vitality utilization, the state and area of IoT gadgets and so forth (Bhatnagar, Shukla, and Majumdar 2019). The utilization of sensors and IoT is vital to moving the world's agribusiness to a progressively gainful and manageable way. Ongoing headways in IoT, WSNs and information and communication technology can possibly address a portion of the ecological, monetary and specialized difficulties. As the quantity of interconnected gadgets keeps on increasing, this creates all the more enormous information with various modalities and spatial and fleeting varieties. Astute preparation and examination of this large information are important to building up a more significant level of information base that outcomes in better dynamic, determining and dependable administration of sensors. An exhaustive survey of the use of various AI calculations on sensor information in the horticultural environment is given in Mekonnen, Namuduri, Burton, Sarwat, and Bhansali (2020). It further discusses a contextual analysis of an IoT-based information-driven ranch model as a coordinated food, vitality and water (FEW) framework. Productively dealing with the water system process has become important to use water stocks because of the absence of water assets around the world. In AlZu'bi, Hawashin, Mujahed, Jararweh, and Gupta (2019) staining trees and intersperses in the dirt have been monitored using sight and sound sensors to distinguish the degree of plant hunger in smart cultivating. They have altered the IoT ideas to draw a motivation towards the vision of "web of multimedia things". The exploration used web of multimedia sensors to improve the water system process. The directed analyses in that work were capable and may be measured in any smart water arrangement framework. IoT has indicated another course of imaginative research in farming space. Being at beginning stage, IoT should be generally tested in order to get broadly applied in different agrarian applications. To pick up knowledge into the best in class of IoT applications in agribusiness and to recognize the framework structure and key advances, a survey was led in Shi *et al.* (2019). They have finished a methodical writing survey of IoT research and organizations in secured horticulture in the course of recent years and assessed the commitments made by various academicians and associations. Chosen references were bunched into three application areas: agriculture, animal cultivating, and food/agrarian item flexibly recognizability. Besides, they have talked about the difficulties in future research possibilities to help new scientists of this area comprehend the flow of advancement of IoT in agribusiness and to propose increasingly novel and imaginative thoughts later on. Smart agrarian detecting has empowered extraordinary points in applications, making it one of the most significant and important frameworks. For outside estate cultivation, the forecast of atmospheric information, for example, temperature, wind speed and moistness, empowers farmer to improve the yield and nature of harvests. Notwithstanding, it is difficult to precisely anticipate atmospheric patterns in light of the fact that the information is unpredictable, nonlinear and contains various segments. A cross-breed profound learning indicator model was proposed in Jin *et al.* (2020), in which an observational mode disintegration of empirical mode decomposition (EMD) strategy was utilized to break down the atmosphere information into fixed segments with various recurrence attributes. At that point a gated intermittent unit is prepared for each gathering sensor as the sub-indicator, and then the outcomes from the gated intermittent unit (GRU) were interpreted to acquire the forecast outcome. The forecast outcomes determined information about temperature, wind speed and others.

13.3 Different tomato leaf diseases

13.3.1 Early blight

This disease is very not unusual on tomato in addition to potato plants. The early blight is particularly due to Alternariasolani, which is basically a fungus. This is found throughout the United States. This first grows on the older parts of the leaves

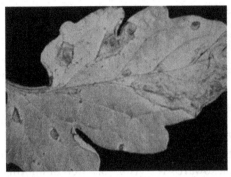

Figure 13.1 Early blight in tomato leaf (Lycopersicon).

Figure 13.2 (a) Green and yellow mosaic pattern on leaf infected with tomato mosaic virus (TMV). (b) Tobacco mosaic virus symptoms on a tomato seedling.

visible as a small brown spot that looks like a "bull's eye". As the disease matures, it spreads outward on the leaf turning it yellow before it dies. Afterwards top part of the plant starts getting infected. This disease is also seen on tomato seeds and tubers of potato. Early blight can develop or appear any time in the course of the growing season. Very high temperatures and wet and humid climatic conditions are most important cause of its spread (Fuentes, Yoon, Youngki, Lee, & Park, 2016) as shown in Figure 13.1.

13.3.2 Mosaic virus

Tomato mosaic virus is one of the oldest plant viruses in tomato plant. It spreads extremely easily and can be destructive to plants, mainly to the tomato. It is hard to find the symptoms of mosaic virus on tomato plant. Tomato mosaic virus symptoms could be found at any stage of growth and all parts may be infected. They are often seen as a general mosaic appearance on surface of the plant. When the plant is severely infected, leaves may become stunted as shown in Figure 13.2.

13.3.3 Target spot

Target spot, or simply we can say early blight, is one of the most common diseases found in the potatoes and tomatoes. It is caused by alternariasolani which is a fungus. It originates as small circles to oval dark brown spots. These spots first enlarge then become oval to angular in shape. During favorable conditions, the individual spots start growing up to a certain height. When the disease becomes severe, all spots unite with each other and cause an upward rolling of the leaf tips and eventually lead to death as shown in Figure 13.3.

13.3.4 Yellow leaf curl virus

Tomato yellow leaf curl is an ailment of tomato plant specifically as a result of a pandemic named tomato yellow leaf curl virus. An infected flower shows the stunted and upright plant growth. Plants are inflamed at an early duration of growth and show severe stunting as shown in Figure 13.4.

Figure 13.3 Close-up to show the ring patterns in the leaf spot caused by target spot on tomato.

Figure 13.4 Yellow leaf curl virus on tomato leaf.

Figure 13.5 Images of diseased tomato leaves from Plant Village Dataset.

13.4 Methodology

For our evaluation (Sankaran, Mishra, Ehsani, & Davis, 2010) as on this, we concentrated only on one plant, tomato. Thus, distinct diseases display exceptional variations, which include shade, shape and illumination in light. In this paper, images have been taken to pick out diseases based on the signs that differentiate one plant from another (Kannan, Prashanth, & Maliyekkal, 2020). The current advancement in the field of machine learning and computer vision requires distinctive features to generate accurate inference. The images used in this work are taken from the internet similar to distinctive types of tomato plants as given in Figure 13.5.

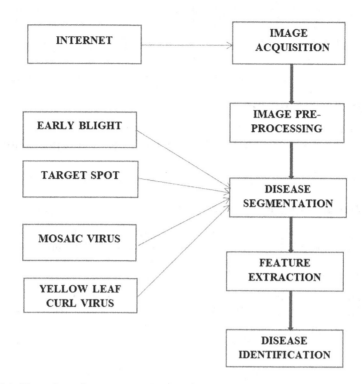

Figure 13.6 Flow chart for tomato plant early disease detection.

The major steps involved in the feature extraction process of tomato leaf images are image acquisition, image pre-processing, segmentation, feature extraction and disease identification as shown in Figure 13.6.

13.4.1 Image acquisition

In this paper, direct image acquisition process was not adopted. The image datasets have been taken from Plant Village Dataset for analysis and evaluation.

13.4.2 Image pre-processing

This process is used to resize the captured images from high to low resolution. Each captured image needs to have a definite length so that it could be analyzed consequently. This intermediate process is necessary to explore accurate statistical characteristic features of the tomato leaves.

13.4.3 Feature extraction

This is the most remarkable step of image processing which helps in dimensional reduction of the image and gives us a compact feature image so that further classification becomes easier for the classifiers (Sankaran *et al.*, 2010). In this paper, following methods have been adopted to explore images: colored histogram and Haar wavelets. The extracted features are used to classify the following diseases: early blight, mosaic virus, target spot and yellow leaf curl virus.

13.5 Disease detection methods

With the help of features extracted earlier, a machine or deep learning based classifier can classify the diseases in tomato plants and an early disease predicator will predict the disease.

13.5.1 Neural network

An artificial neural network (ANN) is a generalized mathematical model which is based on biological nervous systems. The fundamental elements of neural networks are artificial neurons. Input, output and hidden are three basic layers of a simple neural network as presented in Figure 13.7. In feed-forward networks, the data flows from input to output units, firmly in a feed-forward path. Both linear and nonlinear classification problems can be solved by applying ANN with various types of network structures and learning algorithm (Haykin, 2010).

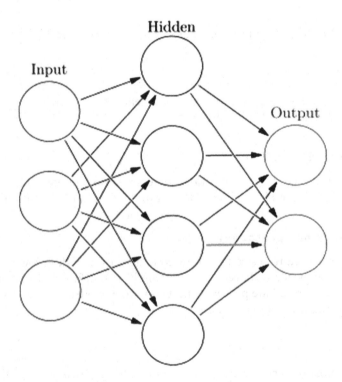

Figure 13.7 Standard ANN structure.

13.5.2 Support vector machine

Among all, neural network–based SVM is one of the most powerful and efficient feed-forward neural networks used for classification and regression problem. It can be used for both linear and nonlinear data classification. It is basically a binary classifier where a nonlinear mapping is considered for transforming the original training data into a higher dimension. The data from one class are separated from another class in this new dimension by using a decision boundary (i.e., a hyperplane). The hyperplane is found by using the support vectors (training tuples) (Beale, Demuth, & Hagan, 1996). In Figure 13.8, the structure of the SVM classifier is presented.

It is explained as follows:

Let D_M be a set of M labeled data points in an N-dimensional hyperspace:

$$D_M = \left[(y_1, a_1), \ldots (y_M, a_M)\right] \in (Y \times A)^M \tag{13.1}$$

where $y_i \in Y$, Y is the input space, and $a_i \in A, A = \{-1, +1\}$.

It is formulated for designing ψ, such that

$\psi: Y \to A$, d is predicted from the input y.

Y can be transformed to an equivalent or high-dimensional feature space to make it linearly separable. The issue of finding a nonlinear decision boundary limit in Y has been mapped to finding an optimal hyperplane for separating two classes.

In the transformed domain, the hyperplane or feature space can be parameterized by (z, c) pair, such that

$$\sum_{i=1}^{Q} z_i \varphi_i(y) + c = 0. \tag{13.2}$$

It is required to calculate the mapping function $\varphi(.)$ explicitly as

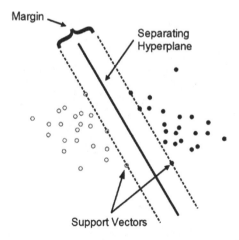

Figure 13.8 Support vector machine.

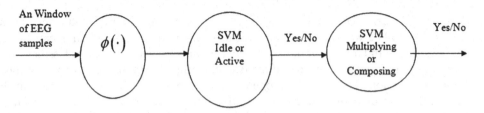

Figure 13.9 Modular SVM structure.

$$\langle \varphi(y_i), \varphi(y_j) \rangle = K(y_i, y_j). \tag{13.3}$$

In the proposed SVM the kernel function is considered as radial basis function. In the input space the patterns after the completion of the transformation are not able to be separated linearly. In Figure 13.9, the structure of the SVM structure is presented.

13.5.3 K-nearest neighbor (KNN) classifier

KNN is one of the most used classification model which is based on analog learning. Here in this method a comparison is done between the training and the testing tuples for finding the similarity. The input to the classifier consists of k-training samples in the feature set. The output of the KNN classifier is a type of class membership. The classification is done depending upon the majority of votes by its neighbors, with the object allocated to the most common neighbors. Here k is a small integer, and when k = 1, then the class is allocated to the class of a single nearest neighbor (Bramer, 2007). The basic structure of the KNN classifier is presented in Figure 13.10.

13.5.4 Random forests

Random forest is a type of classifier which is the combination of multiple classifiers. It works by ensemble learning procedure, and multiple learning mechanisms are used for solving a particular problem. Here, in this method a number of assumptions are constructed and by combining them the problem is solved. Let us consider θ_m is a random vector and free from earlier vectors. The classifier $h(y, \theta_m)$ is generated by completing the training of the data. y is the input data vector in the classifier. After generating a large amount of tree, the voting for most accepted class occurs to get the classification result. The overall structure of this proposed classifier is called random forest, where a group of tree-like another $\{h(y, \theta_m), m = 1, \ldots\}$ is designed for the classification purpose (Mohapatra & Mohanty, 2020). The structure of the random forest model is presented in Figure 13.11.

13.5.5 Deep neural network (DNN)

DNN is a type of advanced neural network based on machine learning model where multiple number of layers are present in the input and output layers. The difference

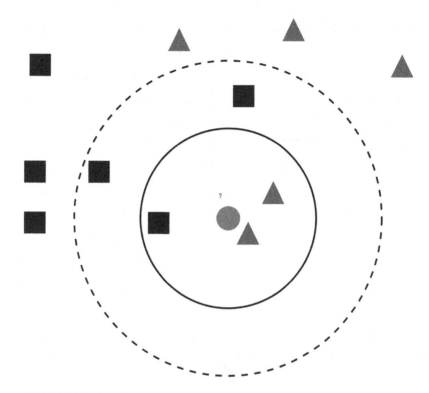

Figure 13.10 KNN classifier.

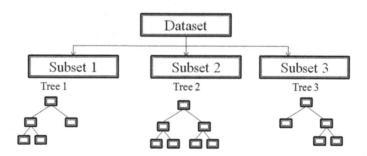

Figure 13.11 Random forest classifier structure.

between a simple neural network and DNN is presented in Figure 13.12. This model is one of the popular artificial neural networks and can be used for different disease classifications. The proposed DNN-based model is designed with two stages. In the first stage, the model automatically learns the features from the input dataset. After successful completion of the feature learning procedure, a fully connected multilayer perceptron classifies the initially learned features. After these two stages, there is a feature identifier module which includes the convolutional and pooling layers. The feature map from the previous layer is convolved by using the convolutional filter (kernel)

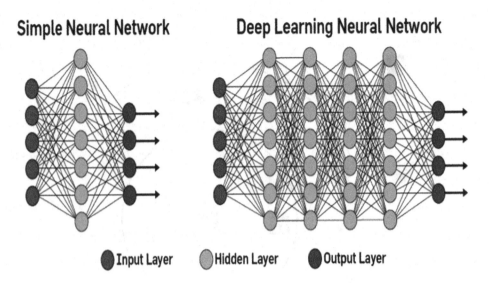

Figure 13.12 Simple neural network versus DNN.

present in the convolutional layer. After completion of the convolution process, it passes through the activation function in order to get the activation map for the next layer. At the same time, the pooling (subsampling) layer creates the activation map to but it increases (Figure 13.12).

In the convolutional layer, the activation map from the past layer is convolved utilizing convolutional channel (or piece) which is included with predisposition and in this way nourished to the actuation capacity to produce an initiation map for the following layer. It is used after the convolutional network. The output of the last layer can be calculated by

$$C_j^{i,a} = \sigma\left(d_a + \sum_{m=1}^{M} w_m^a x_{i+m-1}^{0a}\right),\qquad(13.4)$$

where $x_i^0 = (x_1, x_2, x_3, ..., x_m)$ is the input vector; m, the total number of electrocardiograph (ECG) segments; j, the layer index; and d, the bias of the feature map. Also, σ is the activation function and M the filter size. w_m^a is the weight for mth filter index.

Pooling layer is one of the building blocks of DNN that gradually reduces the size of the activation map to decrease the number of factors and time of computation for the neural network. This layer operates on every feature map separately. Max pooling is one of the common pooling layers used in DNN. The output of a max pooling layer can be found by the maximum activation over a non-overlapping section of input (Mohapatra & Mohanty, 2019; Mohapatra, Srivastava, & Mohanty, 2019; Wu *et al.*, 2018).

Activation function maps an output to a set of inputs. An activation function is generally employed after every convolutional layer. It is a nonlinear transfer function used over the input data. The transfer output is then transmitted to the next layer as the input. Generally, in DNN two types of activation functions are used: (i) Rectified linear unit (Relu), (ii) Softmax.

13.5.5.1 Rectified linear unit (Relu)

It is one of the most used activation functions in DNN. The major advantage behind this activation function is that it does not trigger all the neurons at the same time and converts all the negative input into zero so that the neuron does not get activated. Training performance in this activation function is much faster as compared with other activation functions (Haykin, 1994). It can be represented as

$$f(x) = \max(0, x), \tag{13.5}$$

where x is the input data and f(x) the output function that returns the maximum value between 0 and input data.

13.5.5.2 Softmax

It is a popular activation function as observed from literature. In softmax activation function the exponential of the input signal are considered. Further, the sum of all these values are computed. Next to it the ratio of the exponential to sum of exponential are evaluated as the output function. The advantage of this function is the output probabilities. The range of probabilities varies between 0 and 1. Mathematically softmax activation function can be represented as

$$s_j = \frac{e^{x_j}}{\sum_{i=1}^{n} e^{x_n}}, \tag{13.6}$$

where the input is x and the output value of s is between 0 and 1 and their sum is equal to 1.

13.6 Results and discussion

In this work, 250 tomato leaf images were taken from Plant Village Dataset. Out of which, 50 were the healthy tomato leaf images and 200 were the diseased tomato leaf images. To evaluate the similarities or differences of each disease, we first visualize the histogram of each analyzed image and compare it with a sample of the various diseases in Figure 13.13. Comparing with the features of healthy tomato like mean feature, there is very slight deviation in the values of mean, whereas it is observed that greater variation in mean values was seen when considering diseased tomato leaves. Table 13.1 shows the accuracy of different machine learning models for detecting the diseases in tomato leaves.

DNN machine learning model provides a better result as compared with other models. Due to the presence of more number of hidden layers, DNN performs better. In the proposed leaf disease detection approach, around 97% accuracy is obtained through

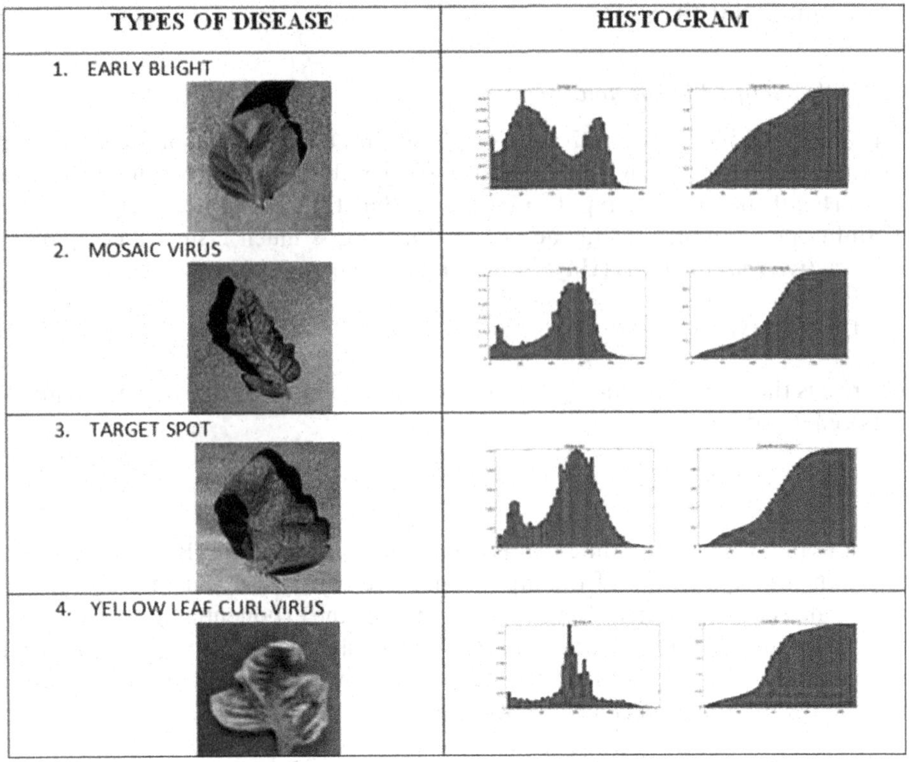

TYPES OF DISEASE	HISTOGRAM
1. EARLY BLIGHT	
2. MOSAIC VIRUS	
3. TARGET SPOT	
4. YELLOW LEAF CURL VIRUS	

Figure 13.13 Histograms of four types of tomato leaf diseases.

Table 13.1 Performance of different classifiers for detecting various tomato leaf diseases

Classifiers	Different tomato leaf diseases and accuracy			
	Early blight (%)	Mosaic virus (%)	Target spot (%)	Yellow leaf curl virus (%)
ANN	88.58	85.69	87.321	86.61
K-NN	82.69	89.58	87.90	87.99
Random forest	79	79.69	87.57	88.06
SVM	90.39	89.98	88.58	87.79
DNN	93.75	91.02	97.59	90.63

DNN model. Also, SVM model performs better as compared with KNN, random forest and ANN.

13.7 Conclusion

In this chapter a brief review is given for application of machine learning in agricultural sector. Also, a case study for detecting various tomato leaf diseases using several machine learning algorithms is presented. As a single case the efficacy of deep learning for detecting the diseases along with comparative results has been shown. So, this chapter may help the researchers in this area for future development of optimal and efficient machine learning model for several agricultural applications.

References

Ale, L., Sheta, A., Li, L., Wang, Y., & Zhang, N. (2019). *Deep learning based plant disease detection for smart agriculture*. Paper presented at the 2019 IEEE Globecom Workshops (GC Wkshps).

Alreshidi, E. (2019). Smart sustainable agriculture (SSA) solution underpinned by Internet of Things (IoT) and artificial intelligence (AI). arXiv preprint arXiv:1906.03106.

AlZu'bi, S., Hawashin, B., Mujahed, M., Jararweh, Y., & Gupta, B. B. (2019). An efficient employment of internet of multimedia things in smart and future agriculture. *Multimedia Tools and Applications*, 78(20), 29581–29605.

Ayaz, M., Ammad-Uddin, M., Sharif, Z., Mansour, A., & Aggoune, E.-H. M. (2019). Internet-of-Things (IoT)-based smart agriculture: toward making the fields talk. *IEEE Access*, 7, 129551–129583.

Beale, H. D., Demuth, H. B., & Hagan, M. (1996). *Neural network design*. PWS, Boston.

Bhagat, M., Kumar, D., & Kumar, D. (2019). *Role of Internet of Things (IoT) in smart farming: A brief survey*. Paper presented at the 2019 Devices for Integrated Circuit (DevIC).

Bhatnagar, A., Shukla, S., & Majumdar, N. (2019). *Machine learning techniques to reduce error in the Internet of Things*. Paper presented at the 2019 9th International Conference on Cloud Computing, Data Science & Engineering (Confluence).

Bramer, M. (2007). *Principles of data mining* (Vol. 180): Springer.

Bu, F., & Wang, X. (2019). A smart agriculture IoT system based on deep reinforcement learning. *Future Generation Computer Systems*, 99, 500–507.

Byabazaire, J., Olariu, C., Taneja, M., & Davy, A. (2019). *Lameness detection as a service: application of machine learning to an internet of cattle*. Paper presented at the 2019 16th IEEE Annual Consumer Communications & Networking Conference (CCNC).

Changmai, T., Gertphol, S., & Chulak, P. (2018). *Smart hydroponic lettuce farm using Internet of Things*. Paper presented at the 2018 10th International Conference on Knowledge and Smart Technology (KST).

Citoni, B., Fioranelli, F., Imran, M. A., & Abbasi, Q. H. (2019). Internet of Things and LoRaWAN-enabled future smart farming. *IEEE Internet of Things Magazine*, 2(4), 14–19.

Das, S. D., Deb, N., Biswal, G. R., & Das, S. (2019). *High voltage aspects of smart agriculture through GIS towards smarter IoT*. Paper presented at the 2019 International Conference on Automation, Computational and Technology Management (ICACTM).

Dharmasena, T., de Silva, R., Abhayasingha, N., & Abeygunawardhana, P. (2019). *Autonomous cloud robotic system for smart agriculture*. Paper presented at the 2019 Moratuwa Engineering Research Conference (MERCon).

Fuentes, A., Yoon, S., Youngki, H., Lee, Y., & Park, D. (2016). *Characteristics of tomato plant diseases—A study for tomato plant disease identification*. Paper presented at the International Symposium on Information Technology Convergence.

Goap, A., Sharma, D., Shukla, A., & Krishna, C. R. (2018). An IoT based smart irrigation management system using machine learning and open source technologies. *Computers and Electronics in Agriculture*, 155, 41–49.

Haykin, S. (1994). *Neural networks: a comprehensive foundation*. Prentice Hall PTR.

Haykin, S. (2010). Neural networks and learning machines, 3/E. Pearson Education India.

Jiang, W., Wang, Y., & Qi, J. (2019). *Study on the integrated model of modern agricultural variety breeding in the Internet of Things environment*. Paper presented at the Proceedings of the 2019 Annual Meeting on Management Engineering.

Jin, X.-B., Yang, N.-X., Wang, X.-Y., Bai, Y.-T., Su, T.-L., & Kong, J.-L. (2020). Hybrid deep learning predictor for smart agriculture sensing based on empirical mode decomposition and gated recurrent unit group model. *Sensors, 20*(5), 1334.

Kannan, U., Prashanth, S. K., & Maliyekkal, S. M. (2020). *Measurement, analysis, and remediation of biological pollutants in water measurement, analysis and remediation of environmental pollutants* (pp. 211–243): Springer.

Katyal, N., & Pandian, B. J. (2020). *A comparative study of conventional and smart farming emerging technologies for agriculture and environment* (pp. 1–8): Springer.

Khanna, A., & Kaur, S. (2019). Evolution of Internet of Things (IoT) and its significant impact in the field of precision agriculture. *Computers and Electronics in Agriculture, 157*, 218–231.

Khoa, T. A., Man, M. M., Nguyen, T.-Y., Nguyen, V., & Nam, N. H. (2019). Smart agriculture using IoT multi-sensors: a novel watering management system. *Journal of Sensor and Actuator Networks, 8*(3), 45.

Kitpo, N., Kugai, Y., Inoue, M., Yokemura, T., & Satomura, S. (2019). *Internet of Things for greenhouse monitoring system using deep learning and bot notification services.* Paper presented at the 2019 IEEE International Conference on Consumer Electronics (ICCE).

Krishnan, R. S., Julie, E. G., Robinson, Y. H., Raja, S., Kumar, R., & Thong, P. H. (2020). Fuzzy logic based smart irrigation system using Internet of Things. *Journal of Cleaner Production, 252*, 119902.

Li, N., Xiao, Y., Shen, L., Xu, Z., Li, B., & Yin, C. (2019). Smart agriculture with an automated IoT-based greenhouse system for local communities. *Advances in Internet of Things, 9*(02), 15.

Mahdavinejad, M. S., Rezvan, M., Barekatain, M., Adibi, P., Barnaghi, P., & Sheth, A. P. (2018). Machine learning for Internet of Things data analysis: a survey. *Digital Communications and Networks, 4*(3), 161–175.

Mekonnen, Y., Namuduri, S., Burton, L., Sarwat, A., & Bhansali, S. (2020). Machine learning techniques in wireless sensor network based precision agriculture. *Journal of the Electrochemical Society, 167*(3), 037522.

Mohapatra, S. K., & Mohanty, M. N. (2019). *Analysis of diabetes for indian ladies using deep neural network cognitive informatics and soft computing* (pp. 267–279): Springer.

Mohapatra, S. K., & Mohanty, M. N. (2020). *Big data analysis and classification of biomedical signal using random forest algorithm new paradigm in decision science and management* (pp. 217–224): Springer.

Mohapatra, S. K., Srivastava, G., & Mohanty, M. N. (2019). *Arrhythmia classification using deep neural network.* Paper presented at the 2019 International Conference on Applied Machine Learning (ICAML).

Mukherji, S. V., Sinha, R., Basak, S., & Kar, S. P. (2019). *Smart agriculture using Internet of Things and MQTT protocol.* Paper presented at the 2019 International Conference on Machine Learning, Big Data, Cloud and Parallel Computing (COMITCon).

Ramdinthara, I. Z., & Bala, P. S. (2019). *A comparative study of IoT technology in precision agriculture.* Paper presented at the 2019 IEEE International Conference on System, Computation, Automation and Networking (ICSCAN).

Salam, A., & Shah, S. (2019). *Internet of Things in smart agriculture: enabling technologies.* Paper presented at the 2019 IEEE 5th World Forum on Internet of Things (WF-IoT).

Sankaran, S., Mishra, A., Ehsani, R., & Davis, C. (2010). A review of advanced techniques for detecting plant diseases. *Computers and Electronics in Agriculture, 72*(1), 1–13.

Shahzadi, R., Tausif, M., Ferzund, J., & Suryani, M. A. (2016). Internet of Things based expert system for smart agriculture. *International Journal of Advanced Computer Science and Applications, 7*(9), 341–350.

Sharma, H., Haque, A., & Jaffery, Z. A. (2019). Maximization of wireless sensor network lifetime using solar energy harvesting for smart agriculture monitoring. *Ad Hoc Networks, 94*, 101966.

Shi, X., An, X., Zhao, Q., Liu, H., Xia, L., Sun, X., & Guo, Y. (2019). State-of-the-art Internet of Things in protected agriculture. *Sensors*, 19(8), 1833.

Valecce, G., Strazzella, S., Radesca, A., & Grieco, L. A. (2019). *Solarfertigation: Internet of Things architecture for smart agriculture.* Paper presented at the 2019 IEEE International Conference on Communications Workshops (ICC Workshops).

Varman, S. A. M., Baskaran, A. R., Aravindh, S., & Prabhu, E. (2017). *Deep learning and IoT for smart agriculture using WSN.* Paper presented at the 2017 IEEE International Conference on Computational Intelligence and Computing Research (ICCIC).

Veloo, K., Kojima, H., Takata, S., Nakamura, M., & Nakajo, H. (2019). *Interactive cultivation system for the future IoT-based agriculture.* Paper presented at the 2019 Seventh International Symposium on Computing and Networking Workshops (CANDARW).

Wu, G., Shao, X., Guo, Z., Chen, Q., Yuan, W., Shi, X., .Xu, Y., & Shibasaki, R. (2018). Automatic building segmentation of aerial imagery using multi-constraint fully convolutional networks. *Remote Sensing*, 10(3), 407.

Yazici, M. T., Basurra, S., & Gaber, M. M. (2018). Edge machine learning: enabling smart Internet of Things applications. B*ig Data and Cognitive Computing*, 2(3), 26.

Yoon, C., Huh, M., Kang, S.-G., Park, J., & Lee, C. (2018). Implement smart farm with IoT technology. Paper presented at the 2018 20th International Conference on Advanced Communication Technology (ICACT).

Zhu, N., Liu, X., Liu, Z., Hu, K., Wang, Y., Tan, J., . . . Jiang, Y. (2018). Deep learning for smart agriculture: concepts, tools, applications, and opportunities. *International Journal of Agricultural and Biological Engineering*, 11(4), 32–44.

Chapter 14

Clock signal and its attribute for agriculture

Abhishek Kumar

LOVELY PROFESSIONAL UNIVERSITY

14.1 Clock-enabled equipment for agriculture

Current farming needs apparatuses and innovations that can improve production effectiveness, product quality, postharvest tasks and diminish their natural effect (Saptasagare and Kodada 2014). Automation in agribusiness contributes to precision farming. Environmental parameters have a persistent effect on the crop from cultivation to cutting. Precision farming accompanies the method of applying the perfect measure of input signals like water, manure, pesticide and so forth at the correct area and at the ideal time to upgrade production and improve quality without affecting the nature. Sensors are helpful agents in modern agriculture; by sensor networks a real-time monitoring system can be developed (Mahato *et al.* 2019). In agriculture, the field conditions like temperature, soil moisture content, water level and pH level of soil can be constantly observed by sensors. A global system for mobile communications (GSM) modem which includes general packet radio services (GPRS) provides information over long transmission range in kilometers. The modem is interfaced with a microcontroller, which sends any unusual conditions like less moisture and rise in temperature, pH concentration and low water level in the tank by means of SMS to farmer's mobile phone and implements the action instructed by the farmer accordingly. Global positioning system (GPS) provides a technique of providing data and information anywhere on the earth. Precise, computerized position tracking with GPS permits ranchers and farmers to naturally record information and apply the same (Reddy 2012). A GPS provides longitude, latitude, altitude and time. Accurate time is necessary for every electronic component. Efficiency is always measured with respect to time (Mahato *et al.* 2019). Every electronics device is equipped by a fine clock signal generated by the oscillator in the range of MHz to GHz frequency. Inversion of frequency provides a time period to complete a task of the system. In this chapter clock signal and their associated properties are discussed in detail.

14.2 What is clock signal?

A clock signal is the reference signal used to compare as a reference in circuits for synchronization of input and output signals. The clock signal has a definite period, and half of the period is known as pulse width which changes its value from high to low or low to high (Jovanović *et al.* 2003). Clock signals are used as the reference or pulse signals in most of the electronic systems today. Many of us generally regard the clock signals as only control signals. However, the following attributes make the clock signals significant (Earl McCune 1994):

1. They possess the greatest fan-out.
2. They travel enormous distances in any circuit.
3. They typically operate at the highest speed possessed in the circuit concerning both control and data signals.

Since the clock signals provide temporal reference for both control and data signals, they must be immaculate, sharp and precise. They should possess high resolution with proper synchronization. On the other hand, the absence of a clock signal can severely limit the operation of the circuit on the whole and even cause catastrophic errors.

The most desirable factor in the excellent performance of digital circuits and mixed-signal VLSI circuits is the synchronization of the system to be able to tackle the very high speed flow of data with fine resolution. With the help of delay in the signals, we can increase not only the time resolution but also the precision that a particular digital signal holds. This is a vital quality in high-tech systems nowadays. Several emerging applications require fine time resolution. Due to the needs like excellent time resolution, clock synchronization and frequency monitoring, we require the delay in the circuits which can be easily achieved with the help of phase-locked loops (PLL) and delay locked loops (DLL) (Johnson *et al.* 1992; Woo 1993).

The areas that require excellent time resolution are (Deanger 1998; Morales 1990; Bult and Wallinga 1988):

1. High-speed communication
2. Military-related security applications
3. Testing and measurement purposes
4. Health-related instruments
5. Accurate control system
6. Agriculture management system

Military and biomedical applications: All the applications like military code encoding and decoding, biomedical sensors as well as patient monitoring systems (PMS) require high precision with high resolution. Hence, the clock generation circuitry in these circuits employs a DLL/PLL with efficient use of delay lines to provide the desirable synchronization. The data sampling rate of these devices is kept faster than the maximum data rate. Thus, they can have the ability to control data timing with precision.

A fine, precise application device requires a clock signal interval as small as 25–600 ps. The clock edge would be lower than the least width by an order multiplying factor. The time reference in the digital devices is responsible for the synchronization of correct arrival of the clock edge and the rate of flow of data on the different clock edge. A few famous examples of digitally synchronized systems are logic analyzers, pulse data generators as shown in Figure 14.1. For this purpose, the synchronization, precision and resolution of the clock signal are of utmost importance. Therefore, the clock signal generation circuitry generally comprises additional elements like buffers and PLL/DLL. PLL and DLL use a variable quality: PLL is based on phase frequency and

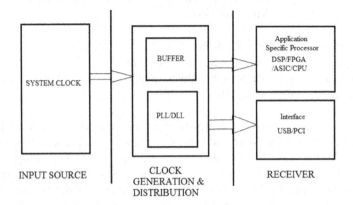

Figure 14.1 Clock signal generation architecture (Santos *et al.* 1996).

DLL is based on variable delay generated by the controlled delay line. This aims to develop programmable delay lines which can easily be configured to achieve the maximum delay. This will be accomplished by adding different kinds of delay units to the fixed delay lines or designing delay lines based upon the usage of different sequential and combinational logic elements.

The clock signal is majorly classified in four categories:

1. Synchronous clock: Each unit of the circuit works on the same frequency. This is the clock frequency generated by oscillator as given to all groups in the design.
2. Asynchronous clock: Each unit in the model receives a different clock; the output of one unit is utilized as a clock reference for the following group. The asynchronous signal can be used for handshake signals for communication and microcontroller.
3. Mesochronous clock: Here, every group is given the same clock frequency, but the phase of the clock is not defined. Recover decides the phase of the aforementioned kinds of signals and a unique technique is required from the data being transmitted.
4. Plesiochronous clock: In this type of clocking strategy frequency at each unit is almost the same, but the phase of the clock is not the same. For every unit, the phase keeps on drifting at a slower speed. Hence, they need too a technique to identify when the active edge of the clock is available on the output drifted from the input by more than half. Delay in the electronic design is defined as the retardation that the signal faces when traveling from source input to destination (Santos *et al.* 1995; Dehng *et al.* 2000; Cheng and Lo 2007).

Different kinds of delays found within the circuit:

1. Propagation delay: It is defined as the time required by the signal to flow from the input terminal to the output terminal of a logic gate. Most frequently it is defined as a rising–falling waveform applied to a block; maximum time is taken from rising input/rising 50% mark to the output rising/falling 50% mark.

2. Gate delay: The maximum time required by a logic gate to compute the expected output. Time taken on receiving an input for generating the expected output is termed as gate delay. For nanometer design, its values are in the unit of nanoseconds with precision picoseconds.
3. Contamination delay: It is the minimum time required from the input 50% threshold to the output 50% threshold. Contamination delay follows the minimum delay along the circuit, that is, it provides the shortest path.
4. Transmission delay: It is the delay that arises due to the data being transmitted which is not dependent on the input or output nodes.

Propagation delay affects most of the circuits and it is by far the largest contributor of all the other delay kinds. Due to delays, the overall speed, performance and effectiveness of the device under test are hampered or sometimes it gives the wrong result. There are various designs in which the paths do not require much effort for having the more precise observations of the speed. However, there are critical paths in the circuit which refer to the longest path in terms of delay. Delay associated with the critical path is known as total delay of the circuit (Kumar *et al.* 2011). Critical paths cannot alter the performance of the overall system and these paths require special monitoring in accordance to the timing details (Singh *et al.* 2012). There are four levels with respect to which the critical paths can be distributed, namely (i) algorithm level, (ii) cell level, (iii) transistor level, (iv) floorplan/layout level.

Standard results are acquired at architecture level. The critical paths are the necessary for determination of total delay that a large circuit will possess. The shortest critical path equalizes the delay exhibited at the output. To reduce the delay, different delay models are incorporated in electron design automation tools at the design level or technology level. The most accessible and effective delay model to compute interconnection and gate delay is RC model. The RC delay model roughly follows the nonlinear transistor C–V and I–V curves with the average value of resistance and capacitance calculated through interpolation and statistical analysis over the switching speed of the gate. This estimation is very effective for delay measurement. Similarly, delay offered through pin capacitance wire and their interconnection is measured using a nonlinear delay model: Elmore delay model and T and Pi RC delay model.

14.3 Delay measurement

During research to understand the delay calculation, an inverter-based delay is the starting point. In the inverter-based delay lines, a minimum delay in the transition of the output from the input is offered. However, delay of the circuit is not only due to delay of inverter gates but also includes the delay offered by interconnection and wires during the transmission of the input to the output. The inverter- and the buffer-based delay lines work in the following manner:

When an input voltage is supplied as the clock signal for reference to the first input of the inverter or the buffer, the input voltage passes through all of the transistors after experiencing some propagation delay and gets transmitted on to the next inverter stage. There can be many inverter stages. The amount of delay achieved increases with the number of stages. To increase the delay of the delay line, we need to exploit the RC delay offered in the circuit. To accomplish this in many designs, a cascaded stage of

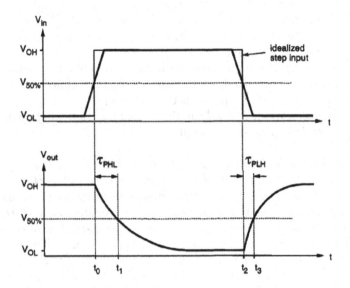

Figure 14.2 Input–output characteristics of NOT gate, τ_{PHL} and τ_{PLH}.

other devices is applied in connection to the main delay line. These devices are called delay elements. The transfer feature of a NOT gate is shown. Input–output signals show variation in the rising and falling curve from the ideal threshold owing to the delay of the inverter circuit shown in Figure 14.2. The propagation delay is divided into two sections: τ_{PHL} and τ_{PLH}. These delays are calculated separately while measuring the delay between the input and output during transitions.

The time taken for the delayed transition between the midway of the rising input and midway of the falling output is given by τ_{PHL}.

The time taken for the delayed transition between the midway the falling input and midway of rising output is given by τ_{PLH}.

The critical assumption made to achieve the previously given switching characteristics of an MOS inverter is that the rise and fall times are zero for each step pulse.

The voltage at $V_{50\%}$ actually is given as

$$V_{50\%} = V_{OL} + \frac{1}{2}\left(V_{OH} - V_{OL}\right) \text{ and}$$

$$V_{50\%} = \frac{1}{2}\left(V_{OH} + V_{OL}\right).$$

Also, from the graph we can evaluate that

$$\tau_{PHL} = t_1 - t_0 \text{ and}$$

$$\tau_{PLH} = t_3 - t_2.$$

Hence, the average value of propagation delay in terms of τ_{PHL} and τ_{PLH} is given as

$$\tau_P = \frac{\tau_{PHL} + \tau_{PLH}}{2}.$$

For the calculation of the values of the delay times, τ_{PHL} and τ_{PLH}, we use the estimate of the charging and discharging average capacitance current. For the capacitance current during a transition being equivalent to I_{avg}, delay times can be represented as

$$\tau_{PHL} = \frac{C_{LOAD}\Delta V_{HL}}{I_{avg,HL}} = \frac{C_{LOAD}(V_{OH} - V_{50\%})}{I_{avg,HL}} \text{ and}$$

$$\tau_{PLH} = \frac{C_{LOAD}\Delta V_{LH}}{I_{avg,LH}} = \frac{C_{LOAD}(V_{50\%} - V_{OL})}{I_{avg,LH}}.$$

Average current while the high to low switching is given as a term dependent upon the current from start to end of a transition period is

$$I_{avg,\ HL} = \frac{1}{2}\Big[i_C\big(V_{in} = V_{OH}, V_{out} = V_{OH}\big) + i_C\big(V_{in} = V_{OH}, V_{out} = V_{50\%}\big)\Big].$$

Similarly,

$$I_{avg,\ LH} = \frac{1}{2}\Big[i_C\big(V_{in} = V_{OL}, V_{out} = V_{50\%}\big) + i_C\big(V_{in} = V_{OL}, V_{out} = V_{OL}\big)\Big].$$

The approach using the average current does not prove to be an efficient method as it ignores the various aspects of current due to capacitance that are present between the input and the output transition. For more accurate results, we solve the state equation for output node in the time domain analysis. This can be achieved by measuring the differential voltage connected with the output terminal which is given below:

$$C_{LOAD}\frac{dV_{out}}{dt} = i_C = i_{D,p} - i_{D,n}.$$

During the conduction period of the NMOS, the initial operation is in the saturation region. However, for the output voltage below $(V_{DD} - V_{T,\ n})$, the NMOS enters the linear region.

Considering the NMOS to be in the saturation region first, we get

$$i_{D,n} = \frac{k_n}{2}\big(V_{OH} - V_{T,n}\big)^2.$$

The saturation region is independent of the output voltage; so, we can approximate the load current as

$$\int_{t=t_0}^{t=t_1'} dt = -\frac{2C_{LOAD}}{k_n\big(V_{OH} - V_{T,n}\big)^2} \int_{V_{OUT}=V_{OH}}^{V_{OUT}=V_{OH}-V_{T,n}} dV_{OUT}.$$

Solving for the time range t_1' to t_0, we get

$$t_1' - t_0 = \frac{2C_{LOAD}V_{T,n}}{k_n\left(V_{OH} - V_{T,n}\right)^2}.$$

For the NMOS in the linear region,

$$i_{D,n} = \frac{k_n}{2}\left[2\left(V_{OH} - V_{T,n}\right)V_{OUT} - V_{OUT}^2\right].$$

Solving between the time interval t_1' and t_1

$$\int_{t=t_1'}^{t=t_1} dt = -2C_{LOAD}\int_{V_{OUT}=V_{OH}-V_{T,n}}^{V_{OUT}=V_{50\%}}\left(\frac{1}{k_n\left[2\left(V_{OH} - V_{T,n}\right)V_{OUT} - V_{OUT}^2\right]}\right)dV_{OUT}.$$

Simplifying, we get

$$t_1' - t_0 = \frac{2C_{LOAD}}{k_n}\frac{1}{2\left(V_{OH} - V_{T,n}\right)}\ln\left(\frac{2\left(V_{OH} - V_{T,n}\right) - V_{50\%}}{V_{50\%}}\right).$$

The final propagation delay times are found by combining Equations (14.1) and (14.2), therefore

$$\tau_{PHL} = \frac{C_{LOAD}}{k_n\left(V_{OH} - V_{T,n}\right)}\left[\frac{2V_{T,n}}{V_{OH} - V_{T,n}} + \ln\left(\frac{4\left(V_{OH} - V_{T,n}\right)}{V_{OH} + V_{OL}} - 1\right)\right].$$

For $V_{OH} = V_{DD}$ and $V_{OL} = 0$,

$$\tau_{PHL} = \frac{C_{LOAD}}{k_n\left(V_{DD} - V_{T,n}\right)}\left[\frac{2V_{T,n}}{V_{DD} - V_{T,n}} + \ln\left(\frac{4\left(V_{DD} - V_{T,n}\right)}{V_{DD}} - 1\right)\right]. \tag{14.1}$$

Similarly, for the charge down event of the capacitance, τ_{PLH} can be found as

$$\tau_{PLH} = \frac{C_{LOAD}}{k_P\left(V_{OH} - V_{OL} - |V_{T,P}|\right)}\left[\frac{2V_{T,p}}{V_{OH} - V_{OL} - |V_{T,P}|} + \ln\left(\frac{2\left(V_{OH} - V_{OL} - |V_{T,P}|\right)}{V_{OH} - V_{50\%}} - 1\right)\right].$$

For $V_{OH} = V_{DD}$ and $V_{OL} = 0$,

$$\tau_{PLH} = \frac{C_{LOAD}}{k_P\left(V_{DD} - |V_{T,P}|\right)}\left[\frac{2V_{T,p}}{V_{DD} - |V_{T,P}|} + \ln\left(\frac{2\left(V_{DD} - |V_{T,P}|\right)}{V_{DD}} - 1\right)\right]. \tag{14.2}$$

14.4 Clock generation

14.4.1 Phase-locked loop

A PLL is an electronic circuit used to produce a signal that maintains the phase in synchronization to the phase reference clock. This device (PLL) consists of a variable frequency generation setup, namely voltage-controlled oscillator, a phase locator, charge pump and clock divider in the feedback path. The phase locator produces a comparison between the input signal and the phase of the reference signal received from the oscillator. The phase locator output is fed into the charge pump circuit, which steps down the charging direction. The production of charge pump goes up for high input and low for low information (Weste and Harris 2011; Prakash and Hiremath 2017; Pahlevan *et al.* 2019). The analog output of the charge pump provides an initial voltage to generate frequency. Frequency of VCO oscillator automatically adjusts the frequency of the oscillator to keep both in phase.

Mathematically, frequency is a derivation of phase. If the input and output phases lock to each other, the input and output frequencies also lock with each other. Hence, a PLL can easily be identified as input frequency and also belongs to the properties to generate any frequency that is a real factor of the input frequency. These features are utilized for modulation–demodulation and synthesis of frequency. The implementation of PLL is done with analog or digital components. The model for the PLL remains the same in both the cases: consisting three components in the forward path and one in feedback path, namely phase-frequency locator (PFD), low-pass filter (LPF), voltage-controlled oscillator (VCO) and charge pump. The accessible structure of PLL is analog PLL (APLL), digital PLL (DPLL), all-digital PLL (ADPLL), software PLL (SPLL), neuronal PLL (NPLL) etc. PLL finds its common applications in radio station, telemetry, FM radio band, telecommunications etc. (Ross *et al.* 2017). PLL is responsible to generate stable clock frequency or clock distribution networks in complex digital designs such as processors. PLL is utilized to produce frequencies from a few hertz to several gigahertz.

14.4.2 Delay-locked loop

DLL is the digital format of PLL, used to insert a delay in the clock signal to the desired value in accordance with the circuit it is included in the design. In usual emphasis, the clock timing properties are delayed, thereby increasing the output produced to an expected level and thus controlled. DLLs are also utilized for clock and data recovery (CDR) systems. To elaborate the functioning of a DLL utilized as a functional block, a negative delay gate is used with the clock reference input signal. It is a kind of PLL circuit only with the difference of the usage of a delay element in place of an oscillator circuit. The internal component of the DLL circuit contains a phase locator (PD) which compares the phase of the reference clock and feedback clock phase. Depending on the UP or DOWN signal of the reference clock, signal charge pump either charges upward or downward. They are shown in Figure 14.3. The charge pump output is filtered out through a low pass filter finally at the far end received by the delay line. Hence, with the phase locator circuitry input, the delay line gets delays in the output clock frequency to the phase locator. The delay line comprises cascaded connections

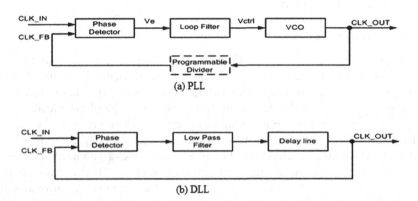

Figure 14.3 PLL and DLL architecture.

of specific delay gates or combinational elements. The delay gates are in the form gates like inverters, NAND gate or the XOR gate.

Another methodology in which the DLL functions is that DLL is applied to the input of the clock which is synchronized with positive (+ve) or negative (–ve) delay. Each unit of the delay in the chain contains a combinatorial multiplexer. The multiplexer control input is kept on updating with another control circuit which generates the (–ve) negative delay effect. Hence, the output of the DLL is positively or negatively delayed clock pulse.

The DLL finds its application in the locking of the clock input with voltage-controlled delay line (VCDL) or digitally controlled delay line (DCDL). The clock signal can be locked up to 1 or 1/2 of the clock cycle when the subtlety increases the duration of lock and it falls behind (Kazemier *et al.* 2017; Casto 2018).

The DLL circuit is not only used to generate fine clock pulse but also performs the function of inserting delay in an existing one. There are three kinds of DLL, namely:

1. Conventional DLL
2. Dual DLL
3. Modified DLL

Delay lines are entirely independent on DLL. Some of the applications are:

1. Low-power devices
2. Wireless transmission architecture
3. Memory elements like SRAM
4. Large processor circuits like microprocessors
5. Handshake signal architecture

The delay can be controlled by a single DC voltage or with the help of multiple binary bits of data.

The DLL differ based on the use of different delay elements. There are three types of delay lines:

1.   Analog or VCDL
2.   Digital delay lines (DDL)
3.   Hybrid delay lines (HDL)

14.5 Controlled delay line

Two basic categories of delay elements for VCDL are inverter- and buffer-based analog VCDL.

14.5.1 Inverter-based VCDL

Inverter-based VCDL device design was started with a schematic design using "Cadence Virtuoso" schematic composer using CMOS 90-nm technology, and simulation result was obtained with cadence spectre simulator. The inverter-based and the buffer-based delay lines work in the following manner: When an input voltage is supplied as the reference edge of the clock to the first input terminal of the inverter or the buffer, the input voltage passes through all of the transistors after experiencing some propagation delay and gets transmitted on to the next inverter stage. There can be many inverter stages. The amount of delay achieved increases with the number of stages. This chapter uses as the basis the thesis work employing a five-stage inverter-based delay line which was made using the Cadence tool. To increase the delay in the delay line, we need to exploit the RC delay offered in the circuit. To accomplish this in practical designs, a cascaded stage of other devices is applied in connection to the main delay line. These devices are called delay elements. Examples of these delay elements include NAND gates, 3T XOR gates.

14.5.2 Buffer-based delay line

A buffer comprises NMOS and a PMOS with the placement of cells in the opposite direction of CMOS inverter. NMOS is kept on the top and PMOS on the bottom, both being the connection of V_{DD} and GND not altered. Here NMOS is connected to V_{DD}, and PMOS is connected to ground. The structure was made such that they only pass the voltage as a buffering unit and do not change the signal. The critical feature of the DLL is that no filtering procedure is required for the input jitter present in the clock path due to noise and unavoidable delay elements. Limitation of DLL is the presence of clock, however, unlike in the PLL, the jitter does not accumulate over the cycles or increase further, it just continues to possess the intrinsic values. The frequency of DLL with application input and output remains the same, whereas the input and output are in delayed format of each other. Due to this, there is no phase error that might get submerged. There are various sources of noise found in DLL circuit like the delay line nervousness to the supply, the device crosstalk and the clock buffers. Hence, a DLL must be designed to have a low jitter input.

14.6 Difference between PLL and DLL

Differences between PLL and DLL are presented as (Gozzelino 2018; Peng 2019; Johnson *et al.* 2016):

1. DLL depends on the variation in phase, but PLL works on the variation in frequency.
2. A DLL does compare the phase of the last stage output with the reference input clock. The DLL produces an error signal which is first assimilated and then feed-back as the control unit to the delay line. This activity makes the zero error which maintains control signal up to down.
3. A PLL is based on the consideration of the phase of the oscillator to produce an error signal. This error signal then assimilates to generate a control signal for the VCO. This control signal reflects its effect on the oscillator's frequency.
4. PLL uses VCO in the feedback, whereas DLL uses a controlled delay line.
5. A clean version of the clock is a significant source of difference between PLL and DLL. PLL blocks the jitter in the source that can affect the VCO output, while DLL propagates the fluctuation.
6. PLL is a viable choice over DLL, where the main sampling point needs to be pulled from the signal that is arriving or ignoring the fluctuation.
7. DLL is used for multiphase sampling delay time to reduce the jitter between two signals.
8. DLL is preferred over PLL at high-speed on-chip communication like communication between DDR SDRAM and memory controller.
9. PLL is a hardware-based frequency control device, while DLL is based on a software dynamic linked library.

14.7 Attribute of clock

Following features measure the quality of clock (Deng *et al.* 2020; Kim *et al.* 2003; Johnson *et al.* 1999, Kim and Kang 2002):

1. Frequency: The term frequency is related to the English word, "frequent", which means to occur many times. Frequency is defined as the number of times a particular event occurs. There are three kinds of frequencies: temporal, spatial and angular frequency. Frequency is reciprocal of the period. Also, frequency is proportional to the inverse of time delay. Thereby, as the delay increases, the frequency decreases. So,

$$f \propto \frac{1}{T} \text{ and}$$

$$f = \frac{1}{2nt_D}.$$

2. Bandwidth: Bandwidth for any signal is defined as the difference between the upper and the lower frequency that a set of frequencies possess. There are different kinds of bandwidths like pass-band or base-band bandwidths. We generally define the pass-band frequency as the difference that is present in between the cut-off frequencies at the upper and the lower end. The amount of information is dependent upon the bandwidth that the signal possesses. Also, depending upon different bandwidths different filters are designed.
3. Operating range: It is defined as the frequency range or limits in which any circuit is operational.

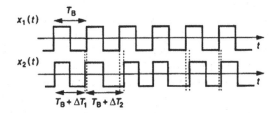

Figure 14.4 Jitter between input clock $x_1(t)$ and $x_2(t)$.

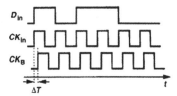

Figure 14.5 Skew generated between data and buffered clock.

4. Jitter: Any periodic waveform that deviates the zero crossings from the original idealistic points is said to be facing fluctuation shown in Figure 14.4.
5. Skew: Generally, in most circuits, only the clock signal suffers from skew. The reason being that they travel the largest distances in any circuit and are constant throughout the circuit. Skew, therefore, is defined as different parts in the circuit receiving the same signal at different time intervals as shown in Figure 14.5. This hampers the proper functioning of the circuit on the whole.
6. Precision: Precision of any system is defined as ability of any system to be able to give the same unrepeated measurements after compiling to the same conditions, and they lead to the same output results as well. The precision of any system is dependent on the system's repeatability and reproducibility.
7. Fine-tuning: Any condition in which the parameters should be absolutely and precisely adjusted so as to agree with the output results.
8. Resolution: The capability of any system to be able to distinguish between a set of control values is defined as the resolution of the system.
9. Coarse grain delay-: Coarse-grain delay can be defined as the delay that is defined at irregular or harsh intervals of time. This kind of delay is not regularly spaced.
10. Fine-grain delay: Fine-grain delay is defined as the delay that can be defined by values occurring over a regular interval of time. The delay is of the appropriate grade.
11. Differential delay calculation: The delay lines cannot be designed for delay elements with an even number of stages because of the problem of locking. The delay elements tend to lock them at a particular value and the signal is not appropriately passed. Therefore, for such delay stages we utilize differential delay calculation, wherein, we add an additional delay element with the help of which we remove the locking error and calculate the delay.

14.8 Conclusion

In this work, we have briefed the clock generation circuit as PLL and DLL. These circuits generate a new clock that can be used as a reference edge in complex devices. Other internal components of the decision are synchronized to the active edge. Mean data transfer is possible on the active edge else the same value of data is to be held. A stable clock is the most important signal of the design. Depending on the application, fine and coarse resolutions are required which overcorrect themself. An important feature of a clock signal is included in the second section of the chapter. The clock is attributed as skew, slew, jitter, etc. which arises due to path delay and interconnect delay in the critical path. Electronic design automation (EDA) tool uses the statistical table to calculate delay as a function of capacitance and input transition. Wire and interconnection delays are caused by the RC delay model with charging and discharging nature of the capacitor. The selection of PLL, DLL and internal component in the path is based on the application.

References

Bult, K., & Wallinga, H. (1988). A CMOS analog continuous-time delay line with adaptive delay-time control. *IEEE Journal of Solid-State Circuits*, 23(3), 759–766.

Casto, M. J. (2018). Multi-Attribute Design for Authentication and Reliability (MADAR) (Doctoral dissertation, The Ohio State University).

Cheng, K. H., & Lo, Y. L. (2007). A fast-lock wide-range delay-locked loop using frequency-range selector for multiphase clock generator. *IEEE Transactions on Circuits and Systems II: Express Briefs*, 54(7), 561–565.

Deanger, J. L. "Digital Delay Line". US PATENT, patent no.5719515, 1998.

Dehng, G. K., Hsu, J. M., Yang, C. Y., & Liu, S. I. (2000). Clock-deskew buffer using a SAR-controlled delay-locked loop. *IEEE Journal of Solid-State Circuits*, 35(8), 1128–1136.

Deng, Z., Liu, Z., Gu, S., Jia, X., & Deng, W. (2020). Frequency-scanning interferometry for depth mapping using the Fabry–Perot cavity as a reference with compensation for nonlinear optical frequency scanning. *Optics Communications*, 455, 124556.

Gozzelino, M., Micalizio, S., Levi, F., Godone, A., & Calosso, C. E. (2018). Reducing cavity-pulling shift in Ramsey-operated compact clocks. *IEEE Transactions on Ultrasonics, Ferroelectrics, and Frequency Control*, 65(7), 1294–1301.

Johnson, A. P., Patranabis, S., Chakraborty, R. S., & Mukhopadhyay, D. (2016, August). Remote dynamic clock reconfiguration-based attacks on the internet of things applications. In *2016 Euromicro Conference on Digital System Design (DSD)*, pp. 431–438. IEEE.

Johnson, C., Demarest, K., Allen, C., Hui, R., Peddanarappagari, K. V., & Zhu, B. (1999). Multiwavelength all-optical clock recovery. *IEEE Photonics Technology Letters*, 11(7), 895–897.

Johnson, M. G., & Hudson, L. "Variable Delay Line Phase Locked Loop Circuit Synchronization System". US PATENT, patent no. 5101117, 1992.

Jovanović, G., & Stojčev, M. K. (2003). Voltage controlled delay line for digital signal. *Facta Universitatis-Series: Electronics and Energetics*, 16(2), 215–232.

Kazemier, J. J., Ouzounis, G. K., & Wilkinson, M. H. (2017, May). Connected morphological attribute filters on distributed memory parallel machines. In *International Symposium on Mathematical Morphology and Its Applications to Signal and Image Processing*, pp. 357–368. Springer, Cham.

Kim, C., & Kang, S. M. (2002). A low-swing clock double-edge triggered flip-flop. *IEEE Journal of Solid-State Circuits*, 37(5), 648–652.

Kim, J., Horowitz, M. A., & Wei, G. Y. (2003). Design of CMOS adaptive-bandwidth PLL/DLLs: A general approach. *IEEE Transactions on Circuits and Systems II: Analog and Digital Signal Processing*, 50(11), 860–869.

Kumar, M., Arya, S. K., & Pandey, S. (2011). Digitally controlled oscillator design with a variable capacitance XOR gate. *Journal of Semiconductors*, 32(10), 105001.

Mahato, S., Rakshit, P., Santra, A., Dan, S., Tiglao, N. C., & Bose, A. (2019). A GNSS-enabled multi-sensor for agricultural applications. *Journal of Information and Optimization Sciences*, 40(8), 1763–1772.

McCune, E. "Binary Controlled Digitally Tapped Delay Line". US PATENT, patent no. 5306971, 1994.

Morales, L. "Active Delay Line Circuit". US PATENT, patent no. 4899071, 1990.

Pahlevan, M., Balakrishna, B., & Obermaisser, R. (2019, May). Simulation framework for clock synchronization in time sensitive networking. *In 2019 IEEE 22nd International Symposium on Real-Time Distributed Computing (ISORC)*, pp. 213–220. IEEE.

Peng, B. (2019). Building a memory reading circuit. *Journal of Computing Sciences in Colleges*, 34(4), 114–116.

Prakash, S. J., & Hiremath, S. S. (2017, August). Dual loop clock duty cycle corrector for the high-speed serial interface. In *2017 International Conference on Smart Technologies for Smart Nation (SmartTechCon)*, pp. 935–939. IEEE.

Reddy, S. R. N. (2012). Design of remote monitoring and control system with automatic irrigation system using GSM-bluetooth. *International Journal of Computer Applications*, 47(12), 6–13.

Ross, B., Carbino, T., & Temple, M. (2017). Home automation simulcasted power line communications network (SPN) discrimination using the wired signal distinct native attribute (WS-DNA). In *Proceedings of the Twelfth International Conference on Cyber Warfare and Security*, pp. 313–322.

Santos, D. M., Dow, S. F., Flasck, J. M., & Levi, M. E. (1996). A CMOS delay-locked loop and sub-nanosecond time-to-digital converter chip. *IEEE Transactions on Nuclear Science*, 43(3), 1717–1719.

Santos, D. M., Dow, S. F., and Levi, M. E. A CMOS delay locked loop and sub nanosecond time to digital converter chip, In *IEEE Nuclear Science Symposium and Medical Imaging Conference*, October 1995.

Saptasagare, V. S., & Kodada, B. B. (2014). Real-time implementation and analysis of crop-field for agriculture management system based on microcontroller with GPRS (M-GPRS) and SMS. *International Journal of Computer Applications*, 975, 8887.

Singh, S., Sharma, T., Sharma, K. G., & Singh, B. P. (2012). New design of low power 3T XOR cell. *International Journal of Computer Engineering & Technology*, 3(1), 76–80.

Weste N. H., & Harris, D. M. *CMOS VLSI Design*. Boston, MA: Addison-Wesley, 4th Edition, 2011.

Woo, A. K. "CMOS Digitally Controlled Delay Gate". US PATENT, patent no. 5227679, 1993.

Printed in the United States
by Baker & Taylor Publisher Services